ref 658.404 6588 £50.00 July 15.

D1429105

Programme Management in Construction

Ali, Kenneth and Peter wish to dedicate this book to Roy.

His passing away is a great loss to his family, friends and the industry.

Programme
Management in
Construction

**Peter T. Barnes, Roy Farren, Ali D. Haidar and
Kenneth P. Wells**

Published by ICE Publishing, One Great George Street, Westminster, London SW1P 3AA

Full details of ICE Publishing sales representatives and distributors can be found at:
www.icevirtuallibrary.com/info/printbooksales

Other titles by ICE Publishing:
Risk Analysis and Management for Projects (RAMP), 3rd edition.
ICE and IFA. ISBN 978-0-7277-4157-8
Joint Ventures in Construction 2.
K Kobayashi, *et al*. ISBN 978-0-7277-5783-8
Art and Practice of Managing Projects.
A Hamilton. ISBN 978-0-7277-3456-3
Planning Major Projects.
RJ Allport. ISBN 978-0-7277-4110-3

www.icevirtuallibrary.com
A catalogue record for this book is available from the British Library

ISBN 978-0-7277-6014-2
© Thomas Telford Limited 2015

ICE Publishing is a division of Thomas Telford Ltd, a wholly-owned subsidiary of the Institution of Civil Engineers (ICE).

Commissioning Editor: Jennifer Saines
Development Editor: Amber Thomas
Production Editor: Rebecca Taylor
Market Development Executive: Elizabeth Hobson

Typeset by Academic + Technical, Bristol
Index created by Nigel d'Auvergne
Printed and bound in Great Britain by TJ International Ltd, Padstow

Contents

Preface

This book imparts the visionary insights and practical experiences needed to understand the multi-faceted nature of programme management today. The book makes good reading for the client, stakeholders, contractors and other practitioners dealing with large projects and programmes in the construction industry worldwide.

The reader is systematically led through broad definitions, to sound philosophy and theory, on to more specific practical implications of the applications of programme management.

The reader is then taken through design team formation and work stages, charter agreements, building information management and on to detailed analysis of current case studies including financial and programme pre-planning illustrations. The scope of services of the programme manager, decision-making tools, planning issues and the different components needed for a successful outcome within a construction programme are illustrated.

Contractual and legal implications are carefully examined, and, finally, insights into future challenges and exciting opportunities facing the well-informed programme manager are explored. Case studies are explained to differentiate how to approach programme management by applying state-of-the-art tools instead of the traditional and well-known project management applications.

About the authors

Peter T. Barnes

Peter T. Barnes is a chartered arbitrator and a UK registered adjudicator, and has an MSc in Construction Law and Arbitration from Kings College, London and a Diploma in International Commercial Arbitration from the Chartered Institute of Arbitrators.

He has been actively involved in the construction and civil engineering industries for almost 40 years, with the last 15 years being spent dealing entirely with claims and dispute avoidance and/or resolution in those industries. Mr. Barnes has acted as arbitrator, adjudicator, mediator and expert witness in respect of both liability and quantum claims.

He is the author of *JCT'05 Standard Building Sub-Contract*, and is the co-author of *Subcontracting Under the JCT 2005 Forms*, *Delay and Disruption Claims in Construction*, and *BIM in Principle and in Practice*.

Peter T. Barnes is based in the UK near London but also operates internationally, and is a director of Blue Sky ADR Ltd, a practice specialising in professional services for the construction and civil engineering industries.

Roy Farren

Roy is a highly creative chartered architect and skilled lead design engineer experienced in delivering complex projects on time and on budget. He has over 20 years' practical design and construction management experience gained in Africa, Europe, the Mediterranean and the Arabian Gulf. He has solid transportation experience, including design for underground rail stations and infrastructure. Roy delivers visionary, yet practical urban design, transportation and environmental solutions for modern communities in the UK, Dubai, Abu Dhabi, Qatar and Saudi Arabia. Familiar with complex regional economic, political and social issues, Roy strives for professional excellence and innovation, delivering successful urban scale projects through consensus and business drive.

Roy is director of Farren Design Consulting, London. He has worked as an architect and urban designer on major urban scale projects in London, Lahore, Dubai, Abu Dhabi, and Durban.

Ali D. Haidar

Dr. Ali Haidar has a PhD in Construction Management, South Bank University, a BSc and an MSc in Civil Engineering from Oklahoma State University and an MSc in Construction Law from Kings College, London.

His experience spans 25 years in the construction industry with much emphasis in recent years on contract management and programme management. He is the author of many papers and apart from his experience in the industry, he has lectured in universities in the UK and the Middle East.

He is the author of three books on claims and he is currently residing in Saudi Arabia and working as a consultant at Dar al Riyadh.

Kenneth P. Wells

Kenneth P. Wells has a bachelor's degree in Building Construction from the University of Florida, USA and has held a certified general contractor's license since 1981 from the State of Florida, USA. The State of Florida calls on Mr Wells as a construction industry subject matter expert for questions to be used in testing for certified specialty contractors.

His experience in the construction industry spans over 33 years, largely in the discipline of project management for large or mega projects and programmes. He has worked in many venues in countries including the United States, the Caribbean and the Middle East.

Mr. Wells is currently residing in Saudi Arabia and working as a consultant at Dar Al Riyadh.

Programme Management in Construction
ISBN 978-0-7277-6014-2

ICE Publishing: All rights reserved
http://dx.doi.org/10.1680/pmic.60142.001

Chapter 1
Introduction to programme management

1.1.   An overview

A programme management team can manage the full scope of work and monitor the full range of activities necessary to keep complex multi-disciplinary projects strictly on time, on budget and to quality specifications undertaken. Programmes ranging from a small number of two or three projects to a large number running into hundreds of projects, often worth billions of pounds, can be managed more effectively than projects managed individually.

Programme management is applied by large organisations, governments and government bodies, usually being the client or acting on behalf of the client or owner, involved in carrying out many projects simultaneously within a defined budget and time to high quality. Programme management is more comprehensive than project management, design management or construction management, as programme management encompasses these three complex domains in a global management operation able to design, manage, execute and control multi-faceted projects simultaneously for one client.

As seen in Figure 1.1, compared to project management, programme management is a relatively new yet developing field, where each project has its own restraints of time, cost and resources and must also be seen in terms of its effect on other projects and resources. It could be said that if project management takes place in the normal three-dimensional world, programme management takes place in a flat two-dimensional world. With building information modelling (BIM), programme management becomes an even more powerful new tool.

As communities grow and government budgets expand to support large and complex programmes, owners, contractors and consultants find themselves responsible for managing multiple complex construction projects and programmes, while balancing limited staff resources, tight schedules and strict budgets.

As the clients' single point of contact, the programme manager integrates the activities of all multi-disciplinary parties (often from diverse backgrounds and cultures) involved in constructing a programme – for example, designers and architects; contractors and subcontractors; and specialist consultancies such as facility management, quality assurance and control, health and safety and environment, interior design, and contract management – to ensure the success of the overall programme.

One of the reasons that clients choose programme management is its ability to provide all the necessary services in house, reducing the need for multiple external consultants. This capability is changing the face of the management of multiple construction projects that are built simultaneously. Fully integrated programme management teams offer services and support in design,

Figure 1.1 Comparison between (a) traditional management systems and (b) programme management

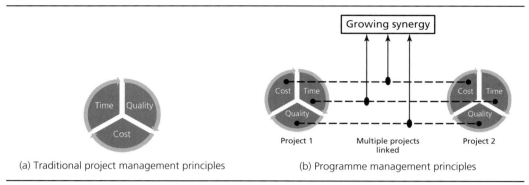

(a) Traditional project management principles   (b) Programme management principles

construction, procurement, project controls, safety, quality and maintenance, thus eliminating delays and lack of coordination of the different activities and enhancing the optimisation of the selection of resources, maximising the efficiency of the construction process of the projects and minimising traditional costs restraints.

Experience, collaborative teams, and proactive communications are essential within a programme management team to make the programme run effectively and on time. When combined, these elements produce high-quality programmes that are technically and environmentally sound, affordable, and completed on schedule.

Programme management generally is responsible for the delivery of the construction of projects through the management of institution relationships; pre-project planning and feasibility; master planning and completion of design documents including conceptual design and design development; project budgets and schedules; procurement of services and specialist providers; administration of contracts; appointment of the contracting team; construction supervision and management; project safety; inspections; furniture, furnishings and equipment; warranty; and project close-out. In many cases the programme manager role extends into the facility management aspect, which for very large programmes can lead to the role of the programme management team extending from the completion date for a further 10 years (Brown, 2008; Martinsuo and Lehtonen, 2006).

1.2. Programme management definitions

Programme management combines the ability and resources to define, plan, implement, and integrate every aspect of the comprehensive programme of multiple projects in the construction industry, irrelevant of their types of construction, from concept to completion, using a team of experts whose sole focus is achieving the client's design and build requirements according to preset key performance indices, milestones, specifications, time and budget.

A programme management procedure can comprise (see Figure 1.2):

■ programme management activities that include planning, monitoring, reporting of ongoing activities and tracking of the cost and schedule; human resources, other administrative support, and the construction of multiple projects in single or multiple geographical locations

Figure 1.2 Programme management procedures and activities

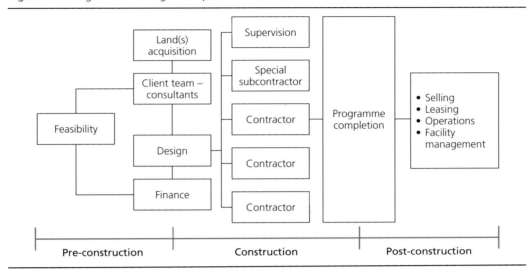

- the process whereby a single entity on behalf of the client exercises centralised authority and responsibility for planning, organising, staffing, controlling, and leading the combined efforts of participating assigned engineers, labour and resources and organisations, for the management of a specific construction-related programme comprising many projects
- the coordinated management of a portfolio of projects to achieve a set of objectives in the construction of a multi-phased programme. In such cases, 'programme' usually refers to an ongoing set of activities within an organisation that create an entity capable of managing one or multiple programmes comprising tens or hundreds of projects simultaneously or within a timeset that runs into years and sometimes into decades
- an understanding of how programmes are designed to use appropriate planning and decision-making strategies to meet programme objectives and aims
- an understanding of how budgets are developed and costs are tracked for individual projects within a programme
- the capability to use indicators and established instruments to document programme performance and outcomes and to plan and optimise the construction of the different projects by allowing resources and money constraints.

Programme management is the process of managing multiple ongoing projects. Examples of these include the construction of 100 schools within a state or country, the rehabilitation of dozens of electrical stations, the construction of tens of military camps or military training grounds and the construction and development of a healthcare programme comprising many hospitals, health centres and clinics. In an organisation, being the client and his or her representatives, and the contractor or the consultant, programme management also involves and emphasises coordinating and prioritising resources across projects, departments and entities to ensure that resource selection is managed from a global perspective.

Programme management can also be defined as the active process of managing multiple global projects which need to meet or exceed business goals according to a pre-determined methodology or lifecycle. Programme management focuses on tight integration, closely knit communications and control exerted over programme resources and priorities.

Programme management is the centralised management to plan, organise, control and deploy key milestones, deliverables and resources from conception through retirement, according to client goals to deliver the programme to a defined design within a budget.

The common threads in these definitions include:

- **Multiple projects**: a programme consists of a series of related and possibly interdependent projects that meet an overarching objective.
- **Planning**: any programme requires complex planning which in theory is similar to project planning, but differs in approach, as will be seen in subsequent chapters. A programme has its own schedule and its own milestones according to key section completion or parts of the programme completion. A programme may entail coordination between and scheduling of a subset of the projects that make up the programme.
- **Monitoring**: the process of management must monitor progress, issues, and risks at the programme level. Programme management entails monitoring by people who have a greater degree of authority than those who are responsible for monitoring in project management.
- **Reporting**: as with monitoring, there must be reporting at the programme level. Programme management consolidates the reports from component projects comprising the programme for its reporting to higher level management.
- **Budget**: in some organisations, projects are responsible for their own budgets but often, the project manager is required to complete tasks and meet deadlines, with budgets that are set at macro levels. Programmes are more often, but not always, inclusive of budget management.
- **Resources**: It is imperative that resources including engineers, labour, equipment, materials and other resources be made available to satisfy the requirements of the projects comprising the programme and the requirements within the time and cost constraints (Project Management Institute, 2008).

The two key characteristic differences between project management and programme management can be summarised thus:

- Programmes encompass a series of projects that in aggregate achieve an overarching set of objectives, whereas projects have specific and more singular objectives. In this sense, the difference is driven by scope and scale.
- Programme management involves more than oversight of a set of projects. It includes application of common standards and processes in the execution of projects, based on a structure that is designed to accomplish the programme (see Figure 1.3).

In summary, programme management addresses the management of project management, setting up processes, and monitoring and measuring project results as well as coordinating related projects and providing the technical, financial support for the delivery of individual projects as well.

1.3.   Programme phases

Programme management in the construction industry is defined as the management of a related series of projects completed over a period of time, to accomplish the complete construction of the individual inter-related projects on time, within budget and according to specifications.

Figure 1.3 Difference between programme management and project management

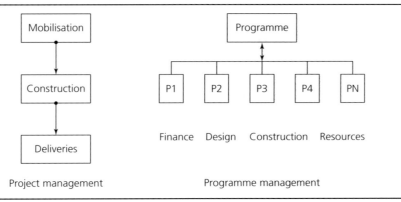

Programme management represents the interests of the client who chooses to oversee their work, to manage directly the construction of their projects, and to coordinate interfaces with other disciplines and stakeholders, by managing their multifaceted projects backed by very sophisticated resources to plan, design, manage and construct a programme. This unique combination provides clients with a one-stop source for the construction of multiple projects.

The following stages (as illustrated in Figure 1.4) are the phases of the construction process for the construction of a programme of multiple projects, and the respective responsibilities of the public entity and the constructor team within that process:

(*a*) **Pre-design phase**: The programme design entity is responsible for the planning, programming, scheduling, budgeting, and financing responsibilities for the construction of multiple projects which it has determined to be necessary and feasible. For an additional scope and fee, the master designer may perform additional services, such as leadership in energy and environmental design (LEED) and quality control and assurances. Architects' design services cover their basic services in the design phase which can vary depending on the scope of work and the size of the complexity of the programme.

(*b*) **Design phase**: The programme manager contracts with architects, engineers, or other consultants to design projects within a programme with definitive plans, and contract documents for bidding and awarding the construction contract to a main large contractor or to a few contractors with resources that are sufficient for constructing the programme.

Figure 1.4 The different steps in programme management

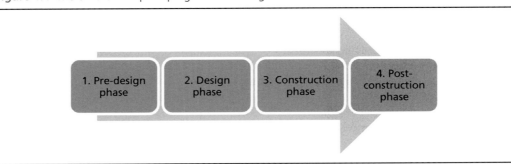

During this phase, the programme manager is responsible for reviewing and analysing the architects' and the engineers' deliverables to help ensure that the designed programme will achieve its goals for cost, schedule, and construction quality.

(*c*) **Construction phase**: The programme management team contracts with a main contractor or contractors to build the projects in accordance with the architect's plans and any consultant-approved modifications. The main contractor then conducts the contract award process for subcontractors, and exercises contract administration. During this phase, the programme manager may appoint the design team or other consultants (for an additional fee) to supervise and administer the work of the general contractor and trade contractors. The client, sometimes, retains responsibility for programme management through the appointment of a supervision consultant service.

(*d*) **Post-construction phase**: Working with the main contractor, the programme manager is responsible for final acceptance, start-up and occupancy of the projects usually done in phases. Sometimes the facility management consultant performs these responsibilities for the programme.

As a general rule, the programme construction phase is sub-divided as follows:

1.3.1   Pre-construction
- Work within the desired delivery system.
- Manage planning and design.
- Evaluate potential sites.
- Assist in selection of design team.
- Maximise front-end planning with early feasibility studies to reduce problems during execution.
- Establish cost and time parameters and prepare bid packages.
- Offer value engineering input and cost analysis.
- Serve as the owner's agent and supplement existing staff.

1.3.2   Construction
- Award contracts.
- Manage construction and coordinate all subcontractor activities.
- Manage material procurement.
- Monitor costs and schedules.
- Maintain quality control.
- Provide ongoing communications and status reports.
- Monitor contractor's safety.

1.3.3   Post-construction
- Develop punch-lists.
- Monitor implementation of punch-lists.
- Resolve outstanding issues.
- Oversee the systems and equipment operations training.
- Remain engaged throughout the warranty period (Ritz and Levy, 2013).

1.3.4   Facility management
The consulting team responsible for the programme facility management must have a deep under-standing of what it takes to deliver cost-effective service improvements and service providers for

the many projects that have been built within the programme. Their involvement must be during the construction stage so that they may get acquainted with the sheer volume and complexity of the works they will undertake.

The team must have experience of working within the full spectrum of facilities management organisations ranging from large outsourced facility management service providers to internal shared services organisations worldwide. Facilities management is a tough commercial environment where efficient service and asset management solutions are vital to maintain healthy margins and customer satisfaction. This becomes a more complex task when facility management has to operate on a large programme running into tens of projects in very different geographical locations and, in some instances, many different types of projects. A programme sometimes can comprise at the same time office buildings, residential compounds, substations, district cooling or district heating plants, sports facilities, educational centres and so on all built at once and delivered at the same time.

It is, therefore, essential that facilities management organisations deploy business processes and systems to manage the complete end-to-end service from initial request to service delivery and final accountability. Facilities management organisations dealing with programmes must work consistently to minimise costs, raise labour productivity and improve services levels enabling them to retain and deliver existing programmes more profitably.

Using different sophisticated methods, facilities management consultants should be able to optimise their services and client satisfaction by:

- providing detailed and accurate information system for documentation, accounting and preventive maintenance for all works including additional unforeseen jobs beyond the scope of the original work orders
- improving cost control of maintenance materials and labour across the programme
- improving efficiency of service delivery through modern notification delivery systems and assignments
- managing multiple projects and sites with many physical locations and providing unique client agreements and rules to define entitlement of services
- increasing the efficiency and effectiveness of service engineers through better manpower planning and scheduling and reducing non-value-added activities.

1.4. The role of the programme manager

A programme manager is often called programme director, programme leader, projects manager, or projects director whose main role is often to oversee multiple project managers who are executing various aspects of the programme of works. Whereas project management is well rehearsed, documented and researched and has well defined criteria which project managers can pursue, programme management is a fairly new management system where the programme manager's range of responsibilities vary from the initial feasibility study of the programme, and the establishment of the scope of work, to the master plan and the comprehensive design and eventually the execution of the projects. Figures 1.5 and 1.6 show the difference in the roles of the project manager and the programme manager.

In summary the programme manager's scope of work is varied and includes such items as:

- the allocation of responsibilities between the various parties involved in the programme
- achieving client objectives
- identifying the contractual matrix of duties between the designers, the main contractors and the various consultants and subcontractors
- establishing budget control
- minimising the client's financial burdens
- imposing quality and safety control for the programme
- executing the complex construction of each project in the programme.

Figure 1.5 The role of a project manager

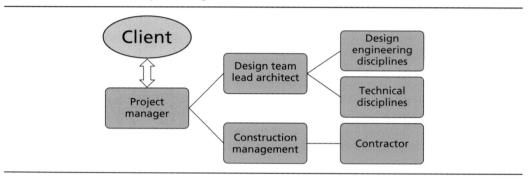

Figure 1.6 The role of a programme manager

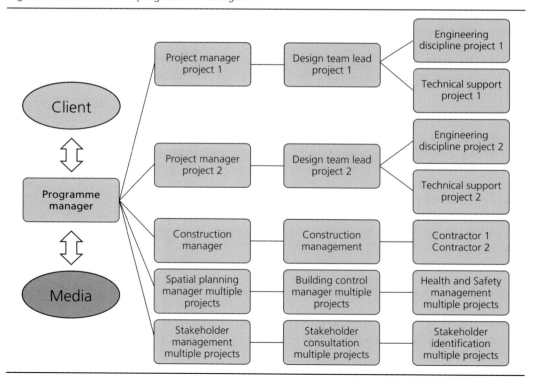

A programme manager, hence, has duties that are more far-reaching and global than those of a project manager. He or she is functioning at a higher level of authority and overseeing many project managers at once, and supervising the procurement process and technical aspects of the programme such as quality, planning, scheduling, quantity surveying methods, shop drawings, reporting, communication, updating and cost control. Hence, a programme manager usually has decades of mega project experience worldwide and usually has a knowledge of all aspects of creating and developing large schemes within the construction industry. He or she should be conversant in design, engineering, accounting, contracts, management systems and structures and with the latest software such as intelligent decision techniques, multi-faceted optimisation, BIM, and, especially, be knowledgeable in facility management and programme handovers.

The services of a programme management team or a programme manager include:

- design and management
- planning and environmental
- construction management
- contracts administration
- change orders, variations and claims management
- procurement
- administrative and financial support
- human resources and participation
- information management
- all-inclusive projects control.

A programme manager's capabilities include experience, resources and expertise in planning, engineering and management. He or she can take an entire programme from first concept through the master plan and initial design; environmental impact study processes and permits; financing and funding; development design, procurement and construction; and start-up and operations, right up to handover and facility management.

A programme manager can manage programmes with full disclosure to and involvement of the owner, as well as full information to the public (as agreed by the owner) and full compliance with all laws, rules, regulations, permits and requirements.

A programme manager has the hands-on experience to develop construction contracts tailored to the needs of the programme. He or she can also support clients with contracting details, including:

- bid preparation for projects
- financing
- marketing
- accounting
- estimating.

Through his or her experience, providing distinctive programme management techniques for clients, a programme manager should have worked with the regulatory agencies of each country where he or she practices. He or she must have developed an in-depth understanding of applicable laws and regulations and their local interpretations.

His or her coordination and/or negotiation services typically include:

- data analysis
- system and structures of large organisations that are client, designer or contractor
- decision-making and analytical skills
- regulatory and public support
- reporting
- risk assessment procedures that allow development to proceed in an innovative, reliable and cost-effective manner without compromising quality or programme requirements.

Although the goal of any construction programme is the finished projects which meet the owner's needs, the programme manager and his or her team have different perspectives and competing interests in the projects. The programme manager wants well-constructed projects which are delivered on time at the lowest possible cost, whereas the constructing team members want to maximise profits in the course of fulfilling their contractual agreements.

Ultimately, the programme manager is responsible for ensuring that the finished projects are within the global budget and meet the needs of the client within relevant environmental conditions, market conditions and time constraints. To assist in managing and monitoring construction projects, the programme manager may enlist the assistance of independent third parties. These parties should represent the interests of the programme manager and exercise oversight independent of the contracting team. This function is also known as construction programme management, or specialist programme management (Tobis and Tobis, 2002).

Once these procedures are accomplished, the programme manager's duties include assisting the client in researching options for projects financing and acquiring suitable locations. Further, his or her duties might include scope to develop a management information control system to monitor the established projects' cost, schedule, financial, and technical requirements throughout the construction process of the programme. Finally, the programme manager could assist the design team's inclusion in the process by integrating the different phases under one main cohesive scope of services. The purpose of the negotiation process is to select an architect that demonstrates that he or she could design a programme within the budgeted cost which could be delivered in a timely manner. The programme manager in the design phase could review and analyse the architect's design in order to produce a clear set of contract documents and specifications for a general contractor or multi-contractors depending on the size of the work and the capacity of the prequalified contractors. To accomplish this goal, the contractor could assist by performing:

- constructability reviews: these reviews help to ensure clear plans and specifications for constructible facilities and to reduce design modifications during the construction of the programme that could lead to higher costs and delays
- lifecycle cost studies: these studies aim to balance the construction cost with the operating and maintenance costs over the anticipated life of the programme in order to provide the construction of the facilities at optimum cost to the public entity
- value engineering studies: these studies evaluate alternative systems, materials, specifications, and construction methods in order to determine the best combination of price, schedule, constructability, function, and aesthetics for each project segment of the programme
- conceptual cost estimates: these estimates for each design continuously monitor and modify the anticipated projects cost of the designed programme

■ problem resolution of design issues: the contractor conducts resolution meetings for design issues between the architect, engineers and other consultants.

A programme manager in the construction phase could perform actions to complete the projects in accordance with goals for cost, schedule, and construction quality. To accomplish this aim, the programme manager could assist the client in the contracting process for the main contractor as well as in the construction administration and management responsibilities for the contractor's construction works. The programme manager could operate the cost accounting and construction cost estimating systems. Further, the programme manager could manage the change order control system for modifications to the approved construction plans, specifications and materials.

Finally, the programme manager could monitor the construction projects for timely completion and resolve issues among the construction team participants.

The programme manager's actions in the post-construction phase complete the construction project. The programme manager could help to ensure that the end users of each finished project within the programme have an operational facility and systems that have been built in accordance with the approved design specifications, which is typical of all different segments of the programme. The programme manager could ensure that the occupant's managing agents have been trained in maintenance and operation of each facility within a set programme. Also, the programme manager might ensure the checking and recording responsibilities for the various construction documents.

Finally, the programme manager could coordinate completion of warranty reviews and post-construction evaluations with the general contractor and the client. The programme manager should ensure that certified individuals have the knowledge and proficiency to perform various tasks in the four phases of the programme process as described in Figure 1.4.

Deep knowledge of architecture, engineering, building science, construction management, or construction technology and extensive experience in managing multi-phased, high-cost construction projects are essential to the role of a programme manager (see Figure 1.7).

Figure 1.7 Typical characteristics of a programme manager

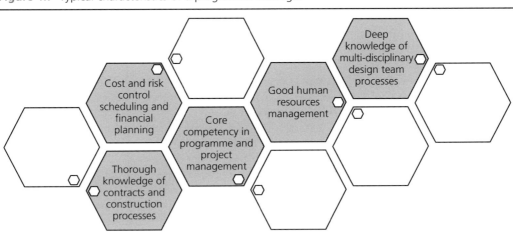

REFERENCES

Brown JT (2008) *The Handbook of Program Management: How to Facilitate Project Success with Optimal Program Management.* McGraw-Hill, New York, NY, USA.

Martinsuo M and Lehtonen P (2006) Program and its initiation in practice: Development program initiation practices in a public consortium. *Proceedings of the 6th European Academy of Management Conference, 17–20 May 2006, Oslo, Norway,* **doi**: 10.1016/j.ijproman.2007.01.011.

Project Management Institute (2008) *The Standard for Program Management, 2nd edn.* Project Management Institute, Philadelphia, PA, USA,

Ritz G and Levy S (2013) *Total Construction Project Management, 2nd edn.* McGraw-Hill, New York, NY, USA.

Tobis I and Tobis M (2002) *Managing Multiple Projects.* McGraw-Hill, New York, NY, USA.

Programme Management in Construction
ISBN 978-0-7277-6014-2

ICE Publishing: All rights reserved
http://dx.doi.org/10.1680/pmic.60142.013

Chapter 2
Programme management concept and scope

2.1. Programme process

Programme process and planning, a distinctive part of programme management, includes for each distinctive project of the programme the individual construction cost estimating, schedule and cost control, projects reporting and review, document control, and structure of the team and organisation. Programme process encompasses all the special management roles such as optimisation, planning, contractual elements such as joint ventures, consortiums and partnerships, and dealing with variation orders and claims management (see Figure 2.1).

Programme process specialists provide support on a project-by-project basis assigned as the specific programme warrants. They are well versed in all commercially available project management techniques which develop proprietary programme management for programmes and special applications for specific projects. They also include mobilisation and systems implementation with the programme management services – initially setting up sophisticated decision-making systems, both hardware and software, and training permanent staff to implement and use the systems to achieve optimal projects management execution.

A common perception is that the current system of managing construction of projects is unsatisfactory because facilities may not be completed on time and within budget or may not meet quality requirements. Several management and oversight solutions are available to address such problems and to help ensure construction of a quality facility that will meet a client's needs and be completed

Figure 2.1 Programme process steps and interrelationships

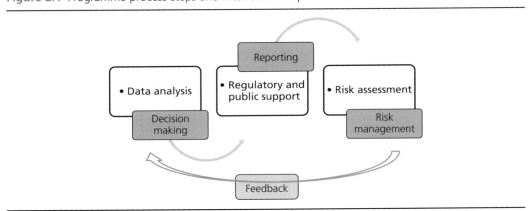

on time. Benefits of using a contractor or many contractors may be compromised by weaknesses in the custom of using ready-made, off-the-shelf standard contracts. The authors found that many standard forms of contracts have deficiencies in their contracting process for many projects, let alone for programmes.

Programme management process always seeks to determine whether it can be clearly defined to include a set of knowledge, skills and abilities which are unique to projects and not possessed or practiced by other professions inside or outside the construction industry; whether specific, objective criteria exist (including measures of cost savings or cost avoidance), which decision makers in the public sector could use to determine the feasibility and benefits of using different contractors; and the extent of current regulatory controls governing the use of programme management (Pellegrinelli, 2008).

Excellence in programme management is achieved through a structured process that includes multiple phases:

■ initiating
■ planning
■ executing, monitoring and controlling
■ completion.

The programme management process balances the key programme constraints and provides a tool for making decisions throughout the programme cycle based on benchmark values, performance metrics, established procedures and the programme aims. Effective programme management includes strategies, tactics and tools for managing the design and construction delivery processes and for controlling key factors to ensure the client receives facilities that match their expectations and function as intended. Improvements in projects quality contribute directly to reduce operational costs and increased satisfaction for the entire programme. Successful projects delivery requires the implementation of management systems that will control changes in the key factors of scope, costs, schedule and quality to maximise the investment, as illustrated in Figure 2.2.

It is critical to establish the qualities of the programme, hence the projects, that are necessary to satisfy the client's and the end user's needs and expectations, and to ensure that they are delivered and in use according to the initial criteria during the early implementation of the programme. Value for money in construction of the programme requires completing projects on time, on budget and to a level of quality that meets the determined needs. A well-planned programme continues to provide value and meet user needs throughout its lifetime and contributes positively to its environment with a wide range of economic benefits.

Figure 2.2 Guidance for the performance of a programme

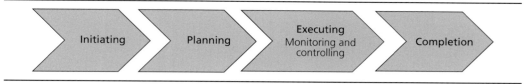

Figure 2.3 Programme delivery: various factors

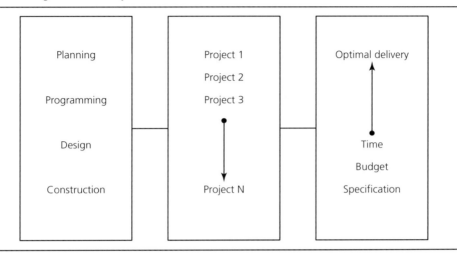

Early investment in planning, programming and design can help deliver these benefits and avoid unnecessary costs and delays (see Figure 2.3). Contemporary institutions and organisations are increasingly realising that traditional methods of management – based on the same approach to every project – cannot meet the needs of today's economic, social and business environment. Additionally, the processes can be streamlined based on technologies and efficiencies not previously available. The responsibility for delivering a programme as planned rests with the entire team. When evaluating options, the whole-life value should be considered and not limited to the short-term initial investment. Factors that affect the longer-term costs of a programme, such as maintainability, useful service life, and resource consumption should be integrated into the decision matrix.

2.2. Programme requirements

Programme inception and preliminary planning require thoughtful definition of goals and needs (programme scope), master planning to accommodate anticipated future needs, evaluation of programme alternatives, identification of sites requirements, funding requirements, budgets for authorisation cycles and/or financial impacts, and programme phasing. There are tools available that help define the goals and objectives for the programme that allow all parties involved to have a decision in making the programme successful. The risks associated with making mistakes in this part of the process are great, since their impact is felt across the programme development process and in the final programme results.

The main parts of the programme requirements, as summarised in Figure 2.4, are:

- scope management
- cost management
- schedule management
- delivery methods
- programme management plans
- design stage management
- delivering and measuring programme quality

Figure 2.4 Programme management components

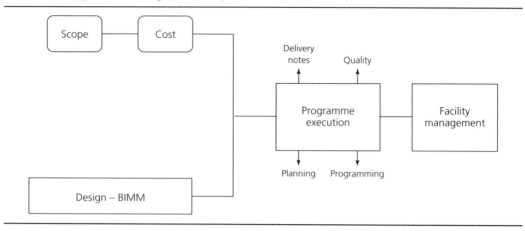

- projects construction stage management
- BIM
- facility management (Dietrich and Lehtonen, 2005).

2.3. Scope management

Programme scope is the work that must be performed to meet a client's programme goals for space, function, features, impact and level of quality. Scope management sets the boundaries for the projects and is the foundation on which the other projects elements are built. From the beginning it helps identify the work tasks and their requirements for completion. Effective programme scope management requires accurate definition of a client's requirements in the planning and development stage of the programme and a systematic process for monitoring and managing all the factors that may impact or change the programme requirements throughout the project design and construction phases through delivery of the completion of each project within the programme.

2.4. Cost management

Programme costs are measured and analysed in certain ways throughout the projects execution milestones, from planning, programming and design to bidding, construction, turnover and post-occupancy (see Figure 2.5). First costs, cost–benefit ratios, and lifecycle costing are a few

Figure 2.5 Cost management components

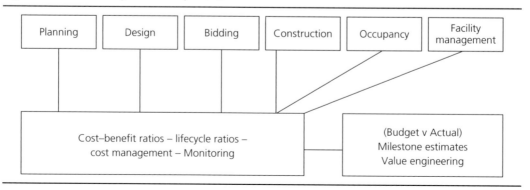

examples of how the programme cost-effectiveness can be evaluated. The control of costs requires continual and systematic cost management and monitoring to compare actual costs incurred against targeted budget numbers. These cost management processes start with the establishment of the initial programme budget based on actual estimates for related projects within the programme and works. They need to align with scope and quality requirements and be based on realistic, current and future market conditions. Comparing budgets to actual costs throughout the programme process is critical. The process continues with milestone estimates, value engineering, procurement strategies and change order management to ensure the programme is timely and cost-effective.

2.5. Schedule management

A programme schedule defines the processes and establishes a timeline for delivering the projects within a programme. Avoiding missing deadlines for delivery of key projects components within the programme is a major objective of schedule management. Comprehensive programme main schedules identify all of the programme stages, phases and activities assigned to each programme manager, mapping them to a timeline that measures key dates that are used to keep track of the programme progress. Schedule management interfaces directly with scope, cost and quality management, and each project member's roles and activities must be defined, coordinated, and continually monitored. It is the goal of every programme manager to look for efficiencies in all of these areas as a programme progresses.

2.6. Programme delivery methods

There are many approaches to achieve successful programme design and construction. The delivery methods are driven by the project's scope, budget and schedule. Some of these methods include traditional (design/bid/build) approaches, integrated delivery process (where all parties have a financial incentive to work together to produce the desired results), EPC (Engineering, Procurement and Construction, mainly used for infrastructure programmes), design–build, bridging, lease–build and lease buy-back. The selection of a proper programme delivery method in turn influences the team composition, schedule, budget and management plans to be followed throughout the process.

2.7. Programme management plan

A programme management plan documents key management and oversight tasks and is updated throughout the programme as changes occur. The plan includes definitions of the client's programme goals, technical requirements, schedules, resources, budgets, management, structure and organisation by establishing sophisticated methods for the client to optimise his resources and to minimise his expenditures. It also provides a vehicle for including efficiencies in the design and construction phases of the programme. It also serves as the basis for completed construction documents and outline the commissioning plan for finished execution of the programme as a whole.

2.8. Design stage management of a programme

Once a design team has been agreed upon and assembled, the client needs to coordinate and manage the programme design phases. Design management requires the oversight of schedules and budgets, review of key submissions and deliverables for compliance with the programme goals and design objectives, verification of the client's input for inclusion, verification of the construction phases, assuring functional testing requirements, and appropriate application of the design standards and criteria. This stage should also define the criteria for assessing quality measurement to ensure the programme success. Determining appropriate goals and objectives at the beginning of the process, and during the master plan and feasibility stages, and measuring their implementation over the lifecycle of construction of the programme have been proven to increase overall programme quality and reduce the programme's accumulative costs and timing to delivery.

2.9.   Delivering and measuring programme quality

A coordinated and well planned design phase is derived from a complex and creative process encompassing a wide range of activities, elements and attributes. Functionality is concerned with the arrangement, quality and interrelationships of the space and locations of each project within the programme and the manner in which the programme is designed, scheduled, planned and constructed. Programme quality relates to the engineering performance of the projects, which includes structural stability and the integration and robustness of systems, finishes and fittings and timely completion within desired budgets. Impact refers to the programme's ability to create a sense of space, function, performance and have a positive effect on the local communities and the environment. Programme quality also encompasses the wider effect the design may have on the art of each project within the programme and architecture. It is the interplay between all of these factors that creates a truly high-performance programme.

2.10.   Programme cost stage management

This stage should include all of the components involved with construction and documentation for the programme. The programme team and the projects' managers involved in this phase are responsible for requests for information, change order management, conflict resolution, inspections, submittal reviews, adhering to schedules and coordinating timely payments. Oversight in this area is critical because it has significant impact on the programme total cost.

2.11.   Building information modelling (BIM)

BIM is the process of generating and managing building data during the programme lifecycle. Typically it uses three-dimensional, real-time, dynamic building modelling software to increase productivity in building design and construction. The process produces the building information model, which encompasses building geometry, spatial relationships, geographic information, and the quantities and properties of the projects' components within the programme. Utilising BIM has the potential to reduce the programme's completion time and overall cost and increase overall productivity of construction and delivery of building projects, with less rework, design and construction errors.

The advantages of BIM over the traditional design and construction process are significant:

(a) BIM single data entry into one model avoids the opportunity for inconsistency and error of repeated input of identical data in multiple media. Data once entered or altered are available in the single-current model available to all.

(b) BIM design efficiency reduces the cost of design and preparing contract documents.

(c) BIM base information is uniform and shared with all participants.

(d) BIM three-dimensionality software identifies physical conflicts between elements, reducing significant construction delay and additional expenses. Where modifications are suggested, the impact of the proposed changes is immediately apparent, subject to evaluation and reconsideration.

(e) BIM assists in sequencing and constructability reviews.

(f) Confidence in shop drawing and fabrication accuracy is improved by BIM because the model can provide construction details and fabrication information. More materials can be fabricated more economically off site under optimal conditions due to confidence in the accuracy of the fabrication.

(g) BIM can link information to quantify materials, size and area estimates, productivity, material costs and related cost information.

Overall, the BIM digital model is a rehearsal of construction and can help identify conflicts and their resolution before actual programme money is spent.

2.12. Quality control

Quality control starts with matching expectations about quality levels within the programme budget and scope during planning and design reviews and continues through the programme construction delivery with a set of inspections, tests and certifications. It requires a coordinated performance among the entire programme team so that a completed programme may fully satisfy users' expectations. Having a process implemented at the beginning of the programme helps to define the goals as well as provide a continuous measurement system over time to make sure the objectives are being met.

2.13. Programme commissioning

Programme commissioning is a quality assurance process for achieving, verifying and documenting that the performance of facility systems and assemblies meet the defined objectives and criteria for the programme. It is a systematic process of ensuring that the programme's building systems perform interactively and effectively according to the design intent and the owner's operational needs. This is achieved by documenting the client's requirements and assuring those requirements are met throughout the entire delivery process of the programme. This involves actual verification of systems performance and integration, comprehensive operation and maintenance documentation, and training of the operating personnel. Programme commissioning services may include commissioning plans, total programme commissioning, systems commissioning, pre-installation performance testing or commissioning, re-commissioning, retro-commissioning, and LEED (Leadership in Energy and Environmental Design) certification. Programme commissioning coordinates and integrates planning development and design decisions and verifies that the delivered facility and its capabilities are efficient and work correctly and that the appropriate training schemes are in place to ensure smooth operations over the facility's life.

2.14. Programme management report

This is a set of reports, documents and published materials that need to be processed, updated and distributed to key personnel within the programme structure. It assures the flow of information, knowledge, number and value of contracts awarded, and is a chronicle of change orders, variations and claims. It is a comprehensive set of documents that preserve the flow of information between all parties involved in a programme. It includes:

- number and value of active projects by type and area
- field engineer workload–project complexity
- number of inspections and reviews made by type and area
- process reviews by phase
- selected emphasis reviews by phase
- a summary of reviews, findings (including frequency and significance), conclusions, recommendations, and disposition or actions taken
- an overall review of accomplishments as they relate to the projects' risk analysis
- programme modifications with supporting explanations.

It also monitors the impact of the construction inspection programme in terms of:

- productivity of reviews
- areas of concern – programme improvements needed or achieved
- adequacy of specifications and plans
- adequacy of construction supervision – manpower management, construction workload
- comments on construction practices attributed to contract documents or bidding practices
- number of documented concerns with resolution
- programme developments, such as materials sampling and testing by contractor, experimental projects and recycling, new methods and equipment, new specifications
- programme cost and time delay trends
- environmental mitigation measures accomplished during the programme construction
- suggested programme changes – programme management, directives, etc.
- use of quality-level analysis
- frequency and documentation of programme contacts
- activities that are not programme-specific
- construction-related promotional activities
- training received by employees and its effectiveness and usefulness
- final assessment of the acceptability of the construction programme
- recommendations for reviews to be considered for future risk analysis.

2.15. Programme management quality assurance

Programme areas where no major problems exist may not require detailed review. As a part of the programme risk assessment, the basis for not making reviews should be documented in the main office files. Programme areas having major problems and those where insufficient information is available for drawing conclusions are candidates to be included in the review cycle.

A fundamental component of construction programme management is an understanding of contract administration and construction quality. Construction quality management involves traditional quality assurance measures employed to control and verify construction, material and product quality. It also encompasses broader topics of continuous quality improvement such as optimisation of decision-making processes, innovative contracting practices for enhancing quality, performance feedback mechanisms, and specification improvements and design refinements.

Quality construction is critical to a successful construction programme. Completed construction projects represent tangible products by which the public measures the success of the programme objectives. The client, being a public or a state entity, ultimately defines the success of construction projects based on the level of delivered quality, which may include a variety of issues such as safety characteristics, operational efficiency during and after construction, materials quality and long-term durability and financial value. The proper use and knowledge of effective con-struction quality management applications, at the programme and the project levels, can produce confidence that completed, government-funded construction work meets the above objectives for success.

Most programme managers are now using some form of statistical quality assurance specifications for their programme construction works. Statistically based specifications are an effective means of ensuring a quality product, and they are a fundamental component of construction quality management. Many programme managers are also using other quality improvement methods, such as obtaining and using projects user feedback, developing performance measures and goals, and using various processes during construction to ensure quality workmanship. All of these quality improvement techniques fall within the broader context of programme quality management.

Quality assurance (QA) is the systematic processes necessary to ensure that the quality of a product is what it should be. Quality assurance is an all-encompassing term that includes quality control, acceptance, independent assurance, dispute resolution, and the use of qualified laboratories and qualified personnel.

All programme structures and management bodies must have a QA system for all types of projects. Construction programme management activities should include elements for encouraging and assisting, implementing or refining their QA plan, and for assessing projects-level implementation of the programme requirements.

2.16. Programme management consultancies' scope of services – a case study

This part of the chapter shows how a programme management consultancy services team might propose itself to a client to get awarded a programme. It shows the focal points of programme management's scope of services. It does not approach the monetary remuneration for programme management, but focuses on the technical merits of the programme manager's role.

2.16.1 Case study (consultant's role)

In its proposal, Programme Management Consultancy Services (PMCS) is presenting its organisation as a world leader in programme construction management. Using innovative planning techniques, PMCS provides services for clients with a group of projects which collectively achieve a beneficial objective for the client or its representative and fall under the auspices of a programme. PMCS will place an emphasis in its planning and programme management techniques on multi projects which often have a long duration and a total construction value that can run into millions of pounds. PMCS, with its integrated programme management services, will provide the organisation with a leadership and strategic planning tool for its complex and multi-project portfolio where the organisation will benefit immensely through the resultant efficiency, harmonisation, planning, scheduling, budget control and quality.

PMCS will provide a management team guiding the overall programme with its dedicated, highly experienced staff that are currently managing construction-related works with a total combined value over UK£60 million (hypothetical number). PMCS will provide clients that concern themselves with every matter that collectively adds to the success of the programme and projects' execution initiative. They will act with direct power from the board, cutting across organisational divisions.

PMCS, through its programme management team, will assist the client as strategic planners, in the projects' implementation by harmonising and prioritising the execution of the projects and their design aspects, in the programme management of the projects, and in the supervision aspects of the projects within the programme. PMCS through its huge experience in the programme's market and its exceptional and large pool of human resources will, as well, be able to provide all clients with all the technical support needed for their programme management consultancy services.

PMCS, through programme management, aims at providing a high-end expertise and knowledge service to successfully deliver multiple large complex projects with one master programme management tool. PMCS, therefore, will present to the client the following benefits:

- The strategy employed by PMCS, through programme management, helps to execute business strategies much more effectively than an uncoordinated approach would do, by ensuring that the programme is executed within the time and costs allocated.
- Through programme management, PMCS will translate strategic business objectives into actionable plans and then manage the tactics to achieve the successful delivery of different projects on multiple locations.
- PMCS will manage and plan projects' cost, time, delivery and quality.
- PMCS will identify global risks and define an early strategy to minimise them.
- PMCS coordinates decision-making within various bodies within an institution if and when required.
- PMCS provides expertise to different sub-organisations within an organisational body.

PMCS will provide the client with a unique integrated programme management tool that can control and plan the execution of all its projects simultaneously with its highly professional staff, its affiliation with some of the best and specialised companies in the world in information technology, planning and scheduling, contracts management, cost control, and quantity surveying, and its long, successful history of managing mega-scale projects worldwide.

2.16.2 Organisation strengths and resources
PMCS is composed of many divisions, with each employing a number of very experienced professionals and executives who form a formidable team and provide highly specialised input for programme management and provide an integrated decision and planning support system for clients embarking on multi projects. PMCS is also affiliated and it has completed many construction projects worldwide with many leading international consultancy firms (see Figure 2.6).

2.16.3 Preparation of detailed design and tender documents
PMCS will oversee and review concept designs and detailed design with tender documents for the projects initiated by the client. The design review will take into consideration all aspects of the design, including coordination with all external utilities. The tender review will comprise reviewing and assessing of the following:

Figure 2.6 Programme management consultancy organisation components

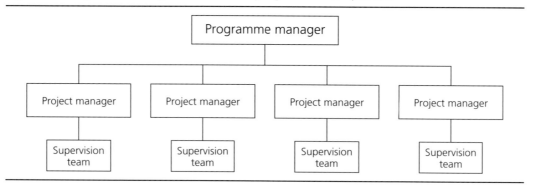

- drawings
- general and special specifications
- bill of quantities and method of measurement
- tender and contract conditions
- cost estimates.

PMCS will oversee the issue of the tender documents to the tenderers, and will assist with managing the tendering process, through the following activities:

- providing answers to tenderers' queries
- participating in the pre-bid conference and site visits, providing clarifications to tenderers' queries, and preparing minutes of the pre-bid meeting
- preparing and issuing to all tenderers addenda to the documents if, during the tender period, the tender documents need to be clarified further
- assisting the client in the receipt of the completed tenders
- taking possession of the completed bid documents for the purpose of tender evaluation.

Other functions will include the following:

- PMCS will establish and agree with the client a framework for tender evaluation and decision-making process. The tender evaluation criteria will be advertised to the tenderers.
- PMCS will, if required, attend the opening of tenders and record details, which will be incorporated in the tender evaluation report.
- PMCS will inspect and review the adequacy and authenticity of all certificates, insurances, bid and performance bonds, indemnities and other legal and official papers for which the contractors are liable under the conditions of contract.
- PMCS will carry out the analysis of the contractual, financial and technical aspects of the tenders received to ensure that they comply with tender documents.

PMCS will also assist the client on the basis of the evaluation, which will take into consideration the following:

- tender–price comparison
- unit–rates comparison
- effect of quantity variations
- proposed construction staff
- proposed construction plant
- compliance with the instructions of tenderers
- work-share between main and sub-contractors
- method of construction
- proposed work plan (see Figure 2.7).

PMCS, if required, will evaluate the results of the analysis and make a recommendation as part of the tender evaluation report and discuss with the client as to which are the best tenders and any matters to be resolved or negotiated before the tenders can be accepted. PMCS will, then, prepare the tender evaluation report, in draft form, for submission to the client. Upon approval, this report will be edited as appropriate and reissued in its final form.

PMCS will, if required by the client, assist in the negotiations with short-listed tenderers and will assemble the final documents incorporating addenda, tender received and documents (if any)

Figure 2.7 Summary of a typical programme management consultancy's scope

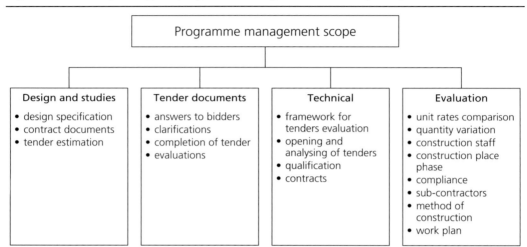

arising from matters negotiated or resolved after tender. PMCS will provide the client with a draft of the award letter and the completed final documents for contract signature.

2.16.4 Programme supervision

PMCS, if requested, will carry out the role of the programme supervision consultant or engineer under the construction contracts with due expedition and without delay. PMCS, if required, can and will provide all supervision works required directly by the client, or else will assist the client by providing the methodology and the preparation of the scope of work and contractual documents.

PMCS's scope shall include, but not be limited to, the following:

- issuing the notice to commence works for the programme
- advising the client on the adequacy of the contractor's insurance policies
- advising the client on the adequacy of the contractor's performance bond and advance payment guarantee
- reviewing the detailed programme construction plan (critical path method or similar) proposed by the contractor to ensure his adherence to contract requirements and client's priorities, instructing, when necessary, the contractor(s) to perform a series of computer runs to introduce modifications to the programme to optimise its results and help the contractor(s) to finalise the same
- establishing and implementing procedures for giving the contractor(s) possession of the sites as stipulated by the contract and in accordance with the agreed plan or otherwise providing advance notice to the client of possible delays due to lack of possession of each project within the programme
- monitoring productivity of labour and plant in relation to the contractor's construction programme as well as material deliveries and consumptions; identifying shortages and informing the contractor of deficiencies in labour, plant or materials, and instructing him to expedite

- monitoring, in particular, the list of long-lead materials which will be required from the contractor
- establishing and enforcing procedures for timely submission of shop drawings and samples to be reviewed and approved
- arranging an initial site meeting for all relevant parties to discuss procedures, means of communication, methods for giving approvals, instructions, variation orders, etc.
- arranging regular site meetings to monitor performance and progress based on the contractor's construction programme and to discuss problems, coordination issues, etc.
- monitoring regularly the construction programme prepared by the contractor and identifying actual or potential delays
- assessing and recommending measures to overcome delays and instructing the contractor accordingly and in accordance with the contract
- reporting to the client and keeping him informed of the estimated completion date.

PMCS, upon the request of the client, will verify that the contractor is constantly complying with the following requirements:

- safety at work: obtain from the contractor and keep a copy on site of the current safety regulations and associated equipment, together with first aid and emergency assistance facilities
- the provision of adequate warning notices, traffic controls and hazard signals to the benefit of the public, operatives and visitors to the sites
- ensuring the compliance of the contractor(s) with the procedural aspects of the programme and the observance of all laws and statutes on nuisance, pollution, etc.

PMCS will prepare and implement systems and forms for processing applications made by the contractor(s) for applications for interim payments, check the site measurements of the quantities of work executed by the contractor(s), and the materials on site, and issue payment certificates as required by the contract (see Figure 2.8).

Figure 2.8 Summary of the programme supervisory roles

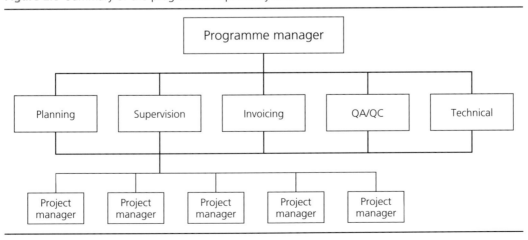

PMCS will submit a monthly progress report to the client, including BIM status reports, which as the digital representation of the physical and the functional characteristics of a macro construction programme comprising multiple projects serve as a shared knowledge resource for information about the programme, forming a reliable basis for decisions during the programme's lifecycle. BIM status reports will include (see Table 2.1):

Table 2.1 Consultant report components

Time control	Delays and their causes, if any
Cost control	Monthly payments and S-curve
Quality control	For site and office submittals
Progress of work	Work done and under progress
Site records	Contractor's site manpower, machinery and weather conditions

2.16.5 Contractual issues

The main contractual issues can be summarised thus:

- PMCS will advise the client on contractual issues as well as matters related to policy, planning and cost control of the programme.
- PMCS, if requested, to issue variation orders to the contractor(s) in accordance with the contract and as required and approved by the client.
- PMCS, if requested, will establish and implement methods and procedures to minimise the potential impact of claims, both financially and time-wise, through prompt and equitable resolution with minimal disruption to the ongoing programme construction activities. This system shall include such activities as receiving, analysing and evaluating claims.
- PMCS will appraise and advise the client on requests for time extensions for different projects within the programme.
- PMCS will assist the client in the settlement of disputes or differences which may arise between the client and the contractor.
- It will advise the client in resolving any disputes or claims from the contractor(s).

2.16.6 Quality control and quality control services

PMCS, through its programme management services, establishes a unified and comprehensive system of maintaining site records for all the projects within the programme, including site correspondence, survey data, inspection records, test data, site diaries, records of meetings, financial records and progress records, by implementing procedures which include the following:

- checking the contractors' setting out
- checking, without relieving the contractors of their obligations under the contract, their method statements and temporary works proposals
- reviewing the submittals made by the contractors for materials, shop drawings and mock-ups in compliance with the contract requirements, by harmonising and standardising the submittal procedure
- arranging the inspection of the construction works for compliance with drawings and specifications and reporting back periodically to the client.

PMCS monitors the quality controls procedures followed by the contractor(s). These procedures cover:

- the inspections and tests carried out by the contractors of all materials and equipment
- the checking by PMCS, as required by the contract, of the competence and suitability of major subcontractors and suppliers the contractor(s) proposes to engage for the different projects within the programme
- the establishment with the contractor(s) of a schedule for testing and commissioning of the installation, witnessing the tests performed by the contractor, and submitting all test reports to the client
- the reception and review of the 'as-built' drawings and instruction and maintenance manuals and submission of copies to the client upon finalisation of the same
- the determination of the date for substantial completion, the preparation of a list of defects and snag items to be rectified and completed during the defects liability period, and the issuing of the taking-over certificate under the contract.

2.17. Design and design activities

Design scope: Design is not always limited to a single project or a single location. In some cases, an owner's requirements may include multiple projects executed over an extended period of time in different locations to fulfil a comprehensive programme according to the client's perspective, objectives and financial aims. For these owners, master planning, prototype design or a combination of the two will be required as an initial step. Private and public owners alike may require these design pre-requisites in anticipation of growing demand for their respective businesses or services.

Master planning: Master planning is the design of multiple projects with integrated and complementary functions, usually on different sites or contiguous sites, to fulfil the requirements of an extensive programme. Master planning allows for future growth or alteration of an owner's facility with minimum obsolescence or loss of function. Master planning requirements should be defined in the owner's programme feasibility study during the early programme conception.

The owner may require design of all or only a portion of a master plan at one time. Accordingly, the design team may be engaged to design all or only a portion of the facilities programme required in a master plan. Types of programmes that normally require master planning include:

- utilities and infrastructures
- schools and universities
- corporate campuses
- manufacturers requiring warehousing, distribution and point-of sale facilities in different locations
- transportation programmes, including ground, rail and airport facilities
- housing units
- defence and military projects.

Master planning design requires thorough knowledge of the owner's present and future requirements and may require more time to complete than an equivalent number of individual project designs of similar extent.

Prototype design: A prototype design is a single design developed for programmes that includes similar projects on different sites. Minor modifications are made to the design to suit site conditions and the requirements of each project. Prototype design establishes a consistent programme identity and purpose that are readily identifiable by facility users. Retail outlets, public buildings, multi-family housing, wastewater treatment plants, bridges, schools and entertainment facilities are examples of programmes executed using prototype design. These kinds of programmes are found in undeveloped countries where there is a huge demand for the development of large multi projects, expanding over many years for infra-structure and social programmes for the development of the wellbeing of citizens. For these kinds of programmes, with multi similar projects, prototype designs can enhance the efficiency of design, construction and facility management stages of the programme's lifecycle.

Conception: At the conclusion of the programme conception stage, a programme delivery method is determined and a design team is selected to perform the task in place. The selection of the design team is critical so the designers are conversant with mastering the design of large and complex programmes. The owner provides the programme and other relevant information to the design team, marking the commencement of the design stage of the programme. From this point forward, a number of participants will become part of the decision-making process, each providing services and information that will shape the character of the programme.

Programme design is the arrangement of programme elements and components, expressed in graphic and written documents, responding to:

- the owner's programme and budget
- requirements of authorities having jurisdiction
- conditions at the different sites where the projects, comprising the programme, will be built
- the owner's functional, aesthetic and sustainability requirements.

Schematic design: In the schematic design phase, the architect or engineer reviews and evaluates the owner's programme and budget requirements and discusses alternative approaches to the design and construction of the programme based on those requirements. As mutually agreed, the design team then prepares schematic design documents for the owner's or owners' representative approval. These may include preliminary sketches, schematic plans, elevations, sections, diagrams, renderings, and other graphic and written documents that illustrate the general extent, scale and relationship of the programme components, and describe in general the type of construction and equipment proposed for each project within the programme.

During schematic design, site plan and area relationships may be defined; the general size, shape and massing of building elements determined; elevations and exterior finishes established; and conceptual design criteria for structural, mechanical and electrical decision making for programme management identified.

Design development: The design development phase follows approval of the schematic design and any necessary programme budget adjustments. The emphasis shifts from overall relationships and functions to more technical issues of constructability and integration of decision making for programme management and components.

Aesthetic concerns move at this stage from massing and arrangement to materials, surfaces and details. Design development phase documents fix and describe the size and character of the entire programme, including architectural, structural, mechanical and electrical details for the whole programme.

Drawings in this phase include plans, elevations and sections that provide more detail of the projects within a programme. The structural, mechanical and electrical drawings for a programme are usually developed by determining preliminary sizes of routing of services. During the development design stage, site plans and floor plans are developed, elevations refined, typical construction details worked out, and many product and material selections made for each project within the programme. Outline specifications are used to record such details and to describe to the owner the materials, products and details for the programme, as well as any special construction conditions or special contract requirements.

2.18. Programme pre-construction documents

The processes of design convert the owner's programme into documents that allow participants to perform their roles, responsibilities and activities related to the programme. During design, the design team, through interaction with other participants, will study the owner's programme and related data produced during the programme conception stage, research the applicable requirements, propose solutions to satisfy the owner's and the programme's requirements, develop renderings and models to communicate the design that best satisfies the requirements, obtain product information, develop graphic and written documents to enhance understanding of the programme by participants, estimate the construction cost of each project within the programme and then estimate the programme construction cost, and obtain the owner's approval at each phase of the design process of the programme (see Figure 2.9).

At the conclusion of the design stage, the participants should have a full understanding of the programme size, and of its functions, components, appearance and probable cost so that the bill

Figure 2.9 Interaction of design phases and pre-construction documents

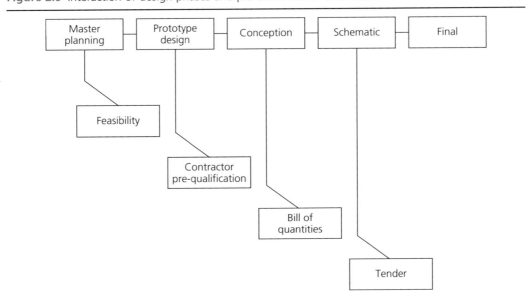

of quantities, the specifications and final design drawings, and full documentation for bidding or negotiation purposes can proceed.

All programmes proceed through multiple stages from conception to facility management, with variations depending on the nature of the work and the needs of the owner.

The graphic and written documents serve as the basis for revised cost projection of the programme, and the outline specifications also serve as a checklist for the subsequent development of detailed documents.

The programme pre-construction documents can be summarised thus:

(*a*) design elements of the programme
(*b*) design development of the programme
(*c*) prequalification and selection of contractors
(*d*) contract formation of the project(s) within the programme
(*e*) bidding
(*f*) bidding analysis and contractor appointment.

2.19.   Programme execution

If the programme is comprised of multi projects, it is essential to allocate the major activities of the projects such as the mobilisation of each project, the substructure, super structure, cladding, block work, mechanical, electrical and plumbing (MEP) finishing and testing and commissioning for a typical building.

In a typical building, the milestones usually are:

(*a*) site development
(*b*) excavation and de-watering
(*c*) foundation and substructure
(*d*) structure
(*e*) block work and plastering
(*f*) MEP
(*g*) cladding
(*h*) lifts
(*i*) finishing.

For a typical programme, the milestones are:

(*a*) land acquisition
(*b*) finance
(*c*) design
(*d*) selection of contractor(s)
(*e*) contract documents
(*f*) execution of project 1 to project n
(*g*) completion and handover
(*i*) occupancy
(*j*) facility management.

2.19.1 Human resources

Each contractor mobilising for multi projects must consider a strong human resources department and administration recruiting engineers and staff. This entity can include a logistics department to coordinate and distribute equipment, materials and labour to the different projects within a programme.

The main functions of this critical entity within a programme management structure consist of:

(*a*) recruiting foremen and labourers for different skills
(*b*) mobilising and distributing the manpower to the different sites
(*c*) providing administrative support such as housing and transportation such as vehicles, buses and plane transport to the different projects, as required
(*d*) organising insurance such as
   (*i*) contractor all risk insurance (CAR)
   (*ii*) contractor plant and machinery insurance
   (*iii*) comprehensive automobile insurance
   (*iv*) all-risks cargo insurance
   (*v*) workmen's compensation insurance.
   (*vi*) health insurance.

2.19.2 Supervision and engineering

The consultant assigned by the client to supervise the works, and in many instances function as the engineer, is vital to the planning of the programme. Many sites cannot be mobilised and do not function without the consultant being present – the consultant must have mobilised his staff with sufficient number and qualified persons for each project.

The supervision consultancy team should be able to arrange and supervise all the activities of the project within a programme. The engineer's role includes but is not limited to:

(*a*) preparing construction cost estimates based on master plan and bills of quantities (if applicable), and establishing overall construction cost budget
(*b*) advising on the cost planning and budget management of the project
(*c*) preparing detailed estimates of availability of schematic design, to be refined progressively with the development of design
(*d*) checking the quotation from sub-contractors and suppliers submitted to the contractor for variations and adding new rates that are not in the priced bills of quantities upon the request of the joint tendering work team consisting of employer and contractor(s)
(*e*) participating in a value engineering exercise initiated by the contractor or the employer to verify the cost effectiveness of engineering
(*f*) reviewing the prices from specialist consultants whose services are commissioned by the contractor and intended to be part of the total construction cost
(*g*) verification by the cost estimator of bills of quantities and other tender documents for infrastructure, of which the corrections and amendments will be carried out by the parties who prepared the bills of quantities. The cost estimator shall prepare unpriced bills of quantities, unit rates, schedule of prices, etc. for all building works within the programme.

2.19.3 Technical support

Centralising the technical support in a multi-project programme is critical to the proper functioning of the projects, as the engineers and the staff on site should be only focusing on the works in situ. The technical support element is usually included within a contractor programme scope management and includes providing the following tasks:

- design shop drawings
- quantity surveying
- invoicing
- planning
- quality assurance
- quality control
- safety
- contract management
- procurement.

In detail, the functions of the technical support department are:

- contractual arrangement with subcontractors
- procurement, which involves the buying and assigning of suppliers and subcontractors
- prequalification of vendors, specialists, suppliers and subcontractors
- providing sub-tender documents including a bill of quantity for the parts necessary – drawings and specifications
- selection of the preferred supplier and subcontractor (not always the best price is chosen as the strength, technical ability, financial capability, workforce and past experience are considered)
- shop drawings – sometimes shop drawings can run into tens of thousands per programme
- programming which ties the activities within each project with the programme.

2.20. Programme management compared to portfolio management

Portfolio management is about all projects, related or unrelated, undertaken by an organisation. These may be divided into functional areas to allow the stakeholders to have a complete view across the organisation in order to understand the various projects that are being carried out.

Often the most useful way of presenting these data is by business function, thereby giving a view of the projects in a particular area. Portfolio management is of interest to decision makers because it gives them a total view of all the initiatives taking place across the organisation. This ensures that the organisation remains focused on what is important, helps avoid duplication and informs strategic decision making. At this level the focus is on the direction of the organisation as a whole, not on individual programmes or projects. The activities undertaken during portfolio management are usually checking strategic alignment and risk management.

For the purposes of this book, programme management is defined as a group of related projects carried out to achieve a defined objective or benefit for a client. Portfolio management, on the other hand, is all projects, which may or may not be connected to one another, being carried out by an organisation (being the client, the contractor or the consultant or engineer).

Programme management is a way to control the management and construction of many projects. It covers vision, aims and objectives, scope, design, approach, resourcing, responsibilities and ultimately the construction of the programme.

As identified earlier, there are four basic stages in programme management. These are programme identification, planning, delivery and handover. These stages take a programme from initiation, based on design, tender, contractual arrangement with contractor(s), supervision, right through to the final construction of a defined programme according to its design and the original vision.

Today most contractors manage multiple projects concurrently with shared or overlapping resources, often in different geographical locations. They may not realise that they are, in fact, building programmes. Traditional project management products and techniques do not recognise the reality of today's contractor's organisational structures and workplace priorities, nor do they leverage the potential benefits that accrue from multi-skilled, multi-location teams.

Programme management is a technique that allows contractors to run multiple related projects concurrently and obtain significant benefits from them as a collection. Programme management is a way to control project management, which traditionally has focused on technical delivery. A group of related projects not managed as a programme are likely to run off course and fail to achieve the desire outcome.

Programme management for a large contractor concentrates on delivering some or all of the following:

■ new capabilities and services
■ business plan
■ strategic objectives
■ change
■ constructability
■ human resources
■ technical support in the form of shop drawings, planning, quantity survey, quality assurance, quality control, health
■ procurement
■ productivity
■ logistics
■ handover.

Programme management and portfolio management can both be described as the management of multiple projects in order to affect major change or gain significant benefits within an organisation, yet the two are significantly different in their scope. A two-fold summary of the differences between programme management and portfolio management follows.

(a) Portfolio management is the coordinated management, construction and delivery of a portfolio of projects to achieve a set of business objectives. Programme management, on the other hand, involves defining the long-term objectives of the contractor. It advises the contractor to set up assorted structures to manage the programme and keep the strategic objectives in mind.

(b) Portfolio management involves the planning and monitoring of tasks and resources across a portfolio of projects. The programme management group within the contractor structure, however, recognises the limitations inherent in methodologies associated with traditional project management procedures and techniques. The programme management group will identify the types of groups of projects and resources that they feel benefit from programme management:

- multiple projects being constructed concurrently
- complex mix of uniquely highly skilled and skilled staff
- geographically dispersed projects
- conflicting priorities and schedules for resources, plants, equipment and projects
- changing deadlines and objectives
- a risk assessment of 'what if' scenarios and requests.

In summary, it is important to recognise the difference between programme management and portfolio management. The term programme management is often used to mean portfolio management and vice versa. The differences between the two, however, include the following:

- Programme management is a group of related projects that are built within one contract and that therefore constitute a programme.
- Portfolio management covers all projects, related or unrelated, being carried out by the contractor.

With this understanding a programme or project manager can differentiate what it has to achieve and how programme and portfolio management fits into the processes and procedures of the client or contractor.

2.21. Summary: objectives of programme management

Programme management is a technique concerned with the construction of a group of related projects, carried out to achieve a defined construction-related objective or benefit for a client, falling under the auspices of a programme (see Figure 2.10). Some projects are simply too large

Figure 2.10 Programme management framework

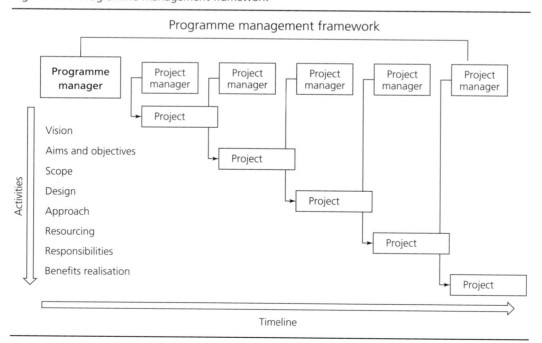

to manage as a single entity: in that case it is necessary to split them up into smaller manageable projects. And if the whole programme is too large for a single project manager to handle, then a number of project managers are required. Smaller projects, thus, are overseen by multiple project managers, all designed to achieve a single long-term objective or benefit for the organisation. In order to control this group and maintain an overall view a programme management system with distinctive entities is required:

- programme
- portfolio
- project
- phase
- strategies.

The programme manager is not concerned with the day-to-day running of individual projects in a programme: this is the project manager's responsibility but he or she needs to ensure that all projects are running on target and that each will achieve its overall contribution to the whole programme. The activities undertaken during programme management are:

- setting the baseline
- agreeing roles and responsibilities
- programme planning
- project prioritisation
- stakeholder communication
- progress reporting
- managing benefits
- quality management
- risk management
- issue management
- programme closure.

REFERENCES

Dietrich, PH and Lehtonen, P (2005) Successful management of strategic intentions through multiple projects – reflections from empirical study. *International Journal of Project Management* **23(5)**: 386–391, **doi**: 10.1016/j.ijproman.2005.03.002.

Eweje J, Turner R and Müller R (2012) Maximizing strategic value from megaprojects: The influence of information-feed on decision-making by the project manager. *Proceedings of the European Academy of Management (EURAM 2011) Conference, International Journal of Project Management*, **30(6)**: 639–651, **doi**: 10.1016/j.ijproman.2012.01.004.

Pellegrinelli S (2008) *Thinking and Acting as a Great Programme Manager*. Palgrave Macmillan, New York, NY, USA, **doi**: 10.1016/j.ijproman.2012.01.004.

Reiss G and Rayner P (2012) *Portfolio and Programme Management Demystified: Managing Multiple Projects Successfully*. E & F Spon, London, **doi**: 10.4324/9780203867730.

Teller J *et al.* (2012) Formalization of project portfolio management: The moderating role of project portfolio complexity. *International Journal of Project Management*, Special Issue on Project Portfolio Management, **30(5)**: 596–607, **doi**: 10.1016/j.ijproman.2012.01.020.

Programme Management in Construction
ISBN 978-0-7277-6014-2

ICE Publishing: All rights reserved
http://dx.doi.org/10.1680/pmic.60142.037

Chapter 3
Pre-planning and decision making

3.1. Pre-planning framework in programme management

The essential purpose of programme management is to provide general control of the objectives of a client's visionary goals. It is achieved through the coordinated management of multiple related projects constituting a programme. Programme management always focuses directly on the strategic business goals and other operational objectives that are defined by the client. Programme managers assist and organise the client and stakeholders in making informed decisions allowing the maximum use of available physical and financial resources. Therefore, programme and project management can be summarised thus:

■ **Programme management** is the active process of managing multiple global construction projects according to a pre-determined methodology or lifecycle. Programme management focuses on tighter integration, closely knit communications and more control over programme resources and priorities to build the programme accordingly (see Figure 3.1).
■ **Project management** is the centralised management to plan, organise, control and deploy key milestones, deliverables and resources from conception through completion, according to the client's objectives for individual project goals. Often project managers have the requisite skills to use specific methods and techniques to manage through the preferred project lifecycle.

Decisions are informed by the programme manager's ability to build credibility, establish rapport and maintain communication with the client's representatives and managers from all levels and entities of the programme structure. The programme manager's main task is to enable the client to clearly view the programme's future direction by effectively managing the issues that are present by making the right decisions at all levels and especially at the early planning stages.

In the pre-planning and pre-construction phase, the first stage is the 'due diligence period'. This stage should be considered the most important first communication rapport established between the programme manager, client, programme team and outside consultants in any programme.

During this period the programme manager defines the key client's needs and objectives, and sets goals and milestones to achieve the completion and the proper constructability of the projects comprising the programme. Those decisions are the backbone of the programme and should be firmly analysed, scrutinised and structured to minimise the associated risks. This is a good time to agree roles, responsibilities and limitations of the client and programme manager.

An example of such a task is the construction of a programme comprising a number of residential units over different geographical areas. This is an example of what the various elements would look like in a programme pre-goal setting. Obviously, the client's vision would be to design, build

Figure 3.1 Client input

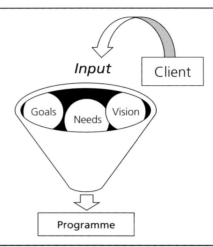

and sell the units within the programme. His goals for sales may very well mirror those highlighted in the feasibility study that was produced for the financial institution which provided the basis for the funding of the programme. Milestone dates at this level would be integrated in the sales pace that is set by the client.

3.2.   Pre-goal-setting

At the time in which both the client and the programme manager have set the goals and milestones, the task of setting policy should be agreed early on as part of the early planning and goal-setting stage. Every programme will need certain policies that are specific to certain requirements of the programme. Some requirements, however, are common for all construction-related programmes. The following list must have policies in place early on to eliminate confusion and errors:

- approval responsibility
- government agency relationships and coordination
- commercial business management
- risk assessment
- key performance indicators (KPI) and performance evaluation
- design criterion
- construction
- planning
- budgeting
- forecasting and scheduling
- quality standards
- safety standards
- purchasing
- execution plan.

Programme policies at the early stages of the programme pre-planning are defined by the stakeholders and owner consultants to make certain that the goals are achieved within the set

Figure 3.2 Cornerstones of planning

parameters. Once the policies are set in place, the procedures can be written to ensure the policies can be realised. Procedures are a series of actions that produce a known result. Checklists are put in place by the programme management office to monitor procedures.

3.3. Execution modelling

No matter who the programme's initiators are – government, commercial or industrial clients – any successful programme has three main cornerstones (depicted in Figure 3.2) that are essential for the pre-planning and the goal setting.

- **Plan of action**: a clear and complete outline that defines the methodology, processes and systems necessary to implement the scope of work of each project within the programme.
- **Integrated time schedule**: the road map to execute and guide the completion of the project or programme deadlines. This schedule should be resource-loaded for all phases of implementation.
- **Budget**: a highly detailed programme budget that is essential to properly track performance during the planning, design and construction phases. Combined with a proper resource-loaded schedule it allows management to predict cashflow issues for the client.

The programme management office, for the pre-planning and goal setting, has five central responsibilities, as listed in Figure 3.3 – namely to manage; to define the scope, schedule and budget; to

Figure 3.3 Central responsibilities

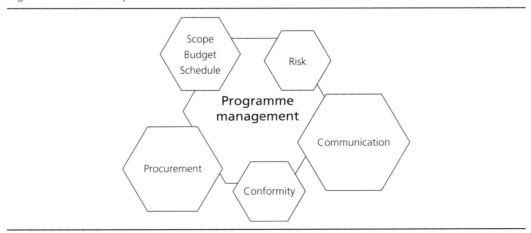

identify the financial, schedule and technical risks; to establish communication; and to provide conformity.

Tasks and deliverables for these responsibilities are as follows:

- **scope schedule pre-budget:**
  - tasks
    - ☐ project identification
    - ☐ priority
    - ☐ analysis
    - ☐ scope selection
  - delivery
    - ☐ estimated cost
    - ☐ schedule baseline
    - ☐ quality standard
    - ☐ milestone dates
    - ☐ funding requirements
- **risks:**
  - tasks
    - ☐ identify financial, schedule and technical threats.
    - ☐ mitigate threats.
  - delivery
    - ☐ risk matrix
    - ☐ gap closure strategy
- **procurement:**
  - tasks
    - ☐ resource sourcing
    - ☐ procurement strategy
    - ☐ procedures economy
  - delivery
    - ☐ standardise resources
    - ☐ milestones incentives
    - ☐ available resources
- **communication**:
  - tasks
    - ☐ status reporting
    - ☐ key performance indicators
  - delivery
    - ☐ dashboard views for stakeholders
    - ☐ programme and portfolio status
    - ☐ earned value analysis
- **conformity**:
  - tasks
    - ☐ phasing
    - ☐ responsibility matrix
    - ☐ process and procedure
    - ☐ management attitude and approach
    - ☐ training staff

- delivery
  - □ templates
  - □ dependencies
  - □ change process
  - □ protocol
  - □ design standards.

Figure 3.4 pulls together the basic concept of the role of the programme manager. Once the vision has been pre-planned with process and procedure (rules) in place, the programme manager's duty is to monitor, advise, direct, redirect, manage and make the right decisions.

3.4. Decision makers' hierarchy in a programme

The project managers share equal responsibility to enact their assigned tasks and roles, thus ensuring the flow of communication. All project managers have the power to prevent information bottlenecks if they prioritise immediate action in coordination with the programme manager. Their role in establishing and agreeing on key decisions, setting up a programme management board and structuring an advisory committee should be paramount.

Project managers should be involved early in the process and make firm and transparent decisions for the benefit of the programme. Their decision is subjective, methodical and involves all risk management analysis. Figure 3.5 represents a simple programme management framework with workflow diagrams.

Figure 3.4 Programme model and interrelationships

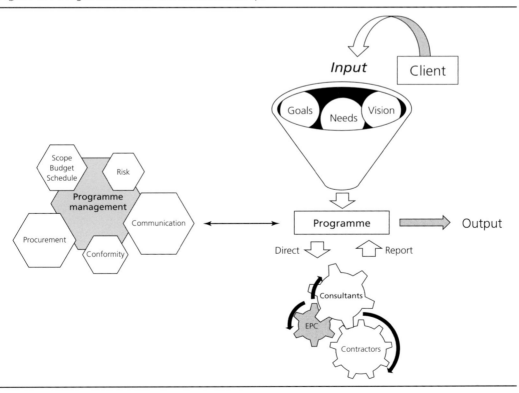

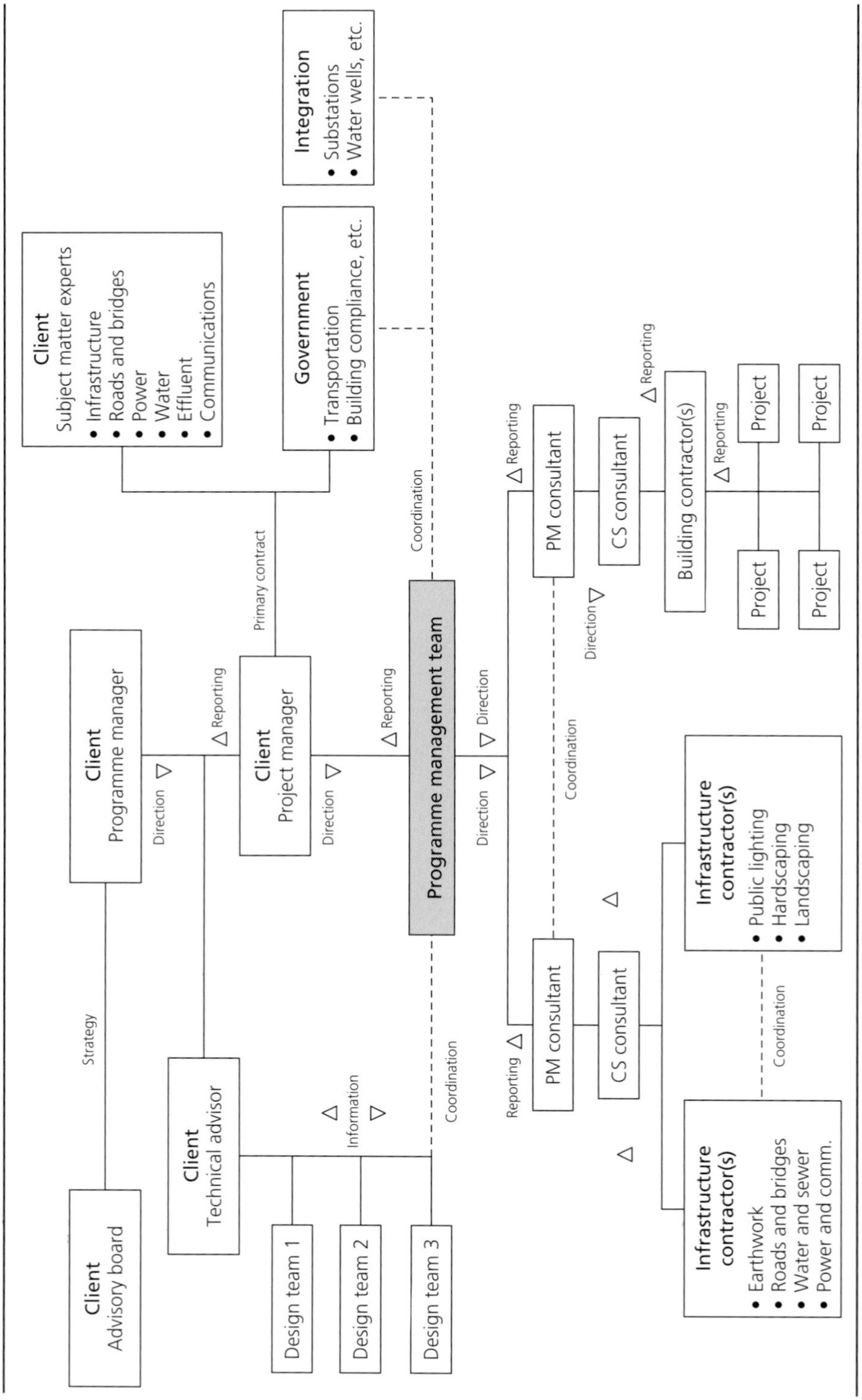

Figure 3.5 Generic programme organisational structure

3.4.1 Programme management team responsibilities during pre-planning

These are as follows:

- **Client**: the owner or client acts to exercise the regulatory and proprietary authority to meet the programme pre-planning goals. The client's board of governors is responsible for providing a clear vision of the pre-planning goals and to make decisions for the benefit of the programme.
- **Government**: the authority responsible for regulatory enforcement of business operations, zoning, code enforcement and any government actions that affect the programme.
- **Programme manager**: the key position in the programme acting as the main conduit of information and direction from the top down to the bottom up. His main tasks in the pre-planning stage are to plan, monitor, analyse and report to the client all decisions. His role also includes decimating information flowing up from the projects, analysing the impact to the programme, reporting to the client and directing the client's decisions down to the projects.
- **Engineers and consultants**: at the pre-planning and goal-setting stage, the consultants, usually the client representatives, shall be involved in all business concerning the design, pre-planning, goal identification and delivery of individual or multiple projects within a programme. The main task shall be to study and establish the rules to monitor, guide, inspect and reports and assure quality assurance and control that will be in place during the construction process.
- **Design consultants**: they are usually involved in the pre-planning and goal identification and setting the architects and designers that are responsible to complete all design and construction documents necessary to convey the magnitude and intent of the scope of work to the owner or client and, possibly, to the contractor. This would include at this stage the master planning and part of the conceptual design.
- **Nominated contractor**: if appointed at this stage, the nominated contractor will be responsible for completing the entire scope of work within the budget and scheduled timeline set by the construction programme contract. The main contractor shall be obligated to assist in forecasting all resources necessary to meet the schedule and to estimate all services in connection with the construction of the programme. Ultimately, his main task is to deliver the assigned programme scope of work, on time and on budget, within the approved parameters. His role is minimal at the pre-planning stage unless the programme is a design and build or EPC programme. Build, operate, transfer (BOT) and public-private partnership (PPP) are also typical examples where the contractor has a greater role in the pre-planning and goal-setting stage interfacing closely with the client.

Figure 3.6 further expands the programme organisational structure of the programme management team indicated in Figure 3.5.

3.5. Decision-making framework

Programme management is a way to control project management. A group of related projects not managed as a programme are likely to run off course and fail to achieve the desire outcome. There are seven key areas in making decisions in a programme framework:

- vision
- aims and objectives
- scope

Figure 3.6 Programme team organisational structure

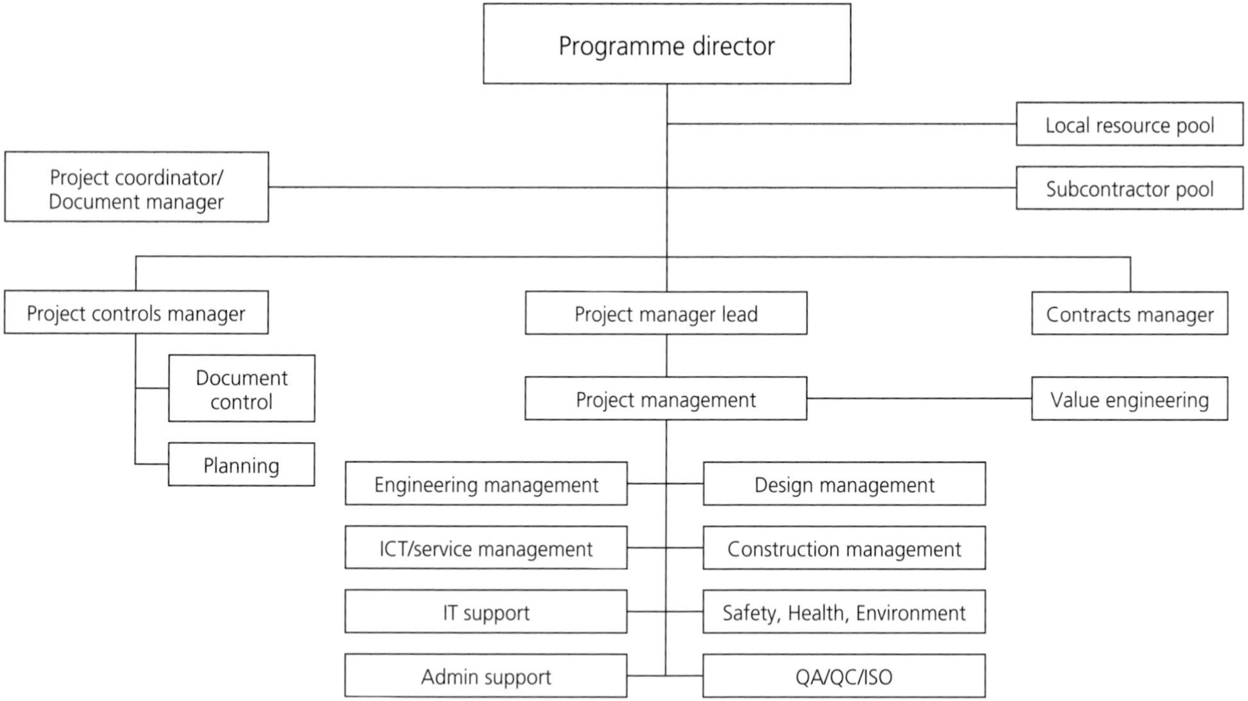

Legend
QA – Quality assurance
QC – Quality control
ISO – International Organization for Standardization
ICT – Information and Communication Technology

- design
- approach
- resourcing
- responsibilities.

Vision is the high-level strategy or idea to drive the organisation towards a goal, benefit or other desired outcome. This vision will usually be a brief statement of intent communicated down from the management or leadership. It is important that the vision has high-level sponsorship and commitment for it to be successful.

The list of aims and objectives is a more detailed statement that explains exactly what is required. It provides a point of reference to go back to when renewed focus is required.

The scope gives boundaries to the programme, explaining what exactly it is that will be delivered; it leaves no room for doubt and means that everyone should be clear about what is being delivered.

The design is the way in which the projects that make up the programme are put together. In this process the programme manager considers which projects have dependencies on others and, therefore, which should come first, which can run concurrently and which should come last.

The approach is the way the programme will be run. The approach is dependent on many factors and it is left to the skill of the programme manager to decide the most effective way. The communication plan is contained within the approach and at the very least should commit to regular progress reporting to the client and the stakeholders.

Resourcing looks at the scheduling and allocation of resources. Short-term and longer-term views should be taken. For the projects in a programme that will start straightaway, it is imperative to identify resources and obtain projects manager's commitment early. For later projects, required resource levels should be identified, but projects managers' commitment is not needed at this stage.

Responsibility identifies and allocates responsibility and risks for each project of the programme. Every member of the programme must clearly understand his or her roles and the roles of the other project managers, designers, finance managers, human resource managers, procurement staff and so on. It is the task of the programme manager to ensure that this is clearly communicated and understood.

The decision-making framework will provide:

- a focus on delivering major organisational changes or benefits
- greater control through visibility of all projects in the programme
- an understanding of project dependencies
- clearly defined roles and responsibilities
- optimised use of resources across projects
- ability to leverage economies of scale and maximise value
- management of risk across related projects
- mechanisms for measuring benefit realisation.

3.6. Programme management decision making

Some of the topics of particular interest to programme management decision-making theory are:

- goals and value decision making for programme managements
- the use of technology and knowledge for a programme
- the structuring of the programme
- formal and informal relationships
- differentiation and integration of activities
- motivation of the programme participants
- group dynamics in programme management
- status and role decision making for programme management in departments within a programme structure
- programme politics
- power, authority and influence in programme management structure
- managerial processes
- programme strategy and tactics
- information decision making for programme management
- stability and innovation in programmes
- programme boundaries and domains
- interface between projects within a programme
- planned change and improvement
- performance and productivity
- satisfaction and quality of work life
- managerial philosophy and programme management culture.

Decision making for programme management is an organised, unitary whole composed of two or more interdependent parts, components or sub-decision-making procedures for programme managers and is delineated by identifiable boundaries from its environment.

3.7. Decision-making components

Programme management decision making consists of a large number of interconnected components. Each component may serve a different function, but the components all have a common purpose. The degree of success in achieving the common goal is a measure of the effectiveness of the decision making for the programme's management. Every decision-making procedure for programme management is either a sub-decision-making procedure of a still larger decision-making procedure for programme management, or a component of the decision making for the programme.

No decision making for programme management is really independent of other decision making for the programme, which simply means there are interactions between the different decision-making processes for the different components of the programme. Any decision making for programme management has a large number of properties. Only some of these properties are relevant to a particular purpose. The values of these properties constitute the state of the decision making for the programme.

The basic components for decision-making contributions are:

- decision making for projects management
- state of a decision-making process

- environment of a decision-making process
- hierarchical decision making for programme management. (A basic concept in decision making for programme managements' thinking is that of hierarchical relationships between decision makers. A decision-making procedure is composed of sub-decision-making procedures for project managers and staff of a lower order and is also part of a supra-decision-making procedure.)
- decision-making analysis
- decision making for programme models.

3.8. The decision-making environment of a programme

Any change in the environment of the decision-making procedure for a programme may affect the whole decision-making apparatus for that programme. The decision-making approach discourages the programme manager from initially presenting a specific problem definition or adapting a particular solution to the problem; rather the decision-making approach for programme managers emphasises that the problematic environment should be defined in broad terms to identify a wide variety of needs that have some relevance to the problem. These needs should reflect the complex relations and conflicts implicit in the problematic environment. Programme decision-making approach covers the comprehensive aspects of the engineering practice and the application of modern decision analysis techniques in the planning and engineering decision making for the programme.

The main focus of decision making for programme management analysis is to optimise the use of resources (people, materials, money and time). Decision-making analysis, thus, involves application of many analytical tools such as utility and theory optimisation, sensitivity analysis, accounting, knowledge base decision making, network techniques, mathematical modelling and, surely, logical and rational behaviour.

Decision making is a significant part of programme management analysis in that:

- It sharpens the programme manager's awareness of the objectives of the project he is designing and planning. The programme manager is required to make explicit statements of what the objectives are.
- It makes precise forecasts.
- It generates broad alternatives.
- It suggests strategies of decision making which can be used to make a selection from possible alternatives.

3.9. Decision-making models

These are abstract representations that describe the interactions between the complex factors of the decision making for the programme environment and the causal dependencies among these factors so that the analysis can correctly perceive the effects of the substantial changes that may be introduced by a large-scale project. The types of model vary:

- iconic (physical presentation)
- analogue (schematic)
- mathematical or analytical
- computer simulation (e.g. Monte Carlo and fuzzy logic)
- artificial intelligence (e.g. knowledge base systems, genetic algorithms and neural networks).

The decision-making model-building process mainly consists of:

- model formulation
- model verification (existing data)
- model application to predict new observations
- model refinement to achieve precision.

The fundamental steps in decision-making modulations are:

- problem definition and statement of objectives
- formulation of measures of effectiveness (MOE)
- generation of alternative solutions
- evaluation of alternatives
- selection and implementation of the decision
- feedback.

3.10.   Contingency view

The contingency view depends on a body of knowledge and research tasks that focus on inter-relationships among key variables and decision making for programme managers. It also emphasises the role of the manager as diagnostician, pragmatist and artist. In terms of decision making for a programme management model, the contingency view suggests sub-decision making is delineated by identifiable boundaries from its environment. The contingency view seeks to understand the interrelationships within and among sub-decision making between the programme and its environment and to define patterns of relationships or configurations of variables. It emphasises the multivariate nature of programme management and attempts to understand how programme management operates under varying conditions and in specific circumstances. This approach recognises the complexity involved in modern programme management but uses the existing body of knowledge to relate environment and design, to match structure and technology, to integrate strategy and tactics, or to determine the appropriate degree of subordinate participation in decision making, given a specific situation. Success in the art of management depends on a reasonable success rate for actions taken in a probabilistic environment.

Contingency views represent a middle ground between the view that there are universal principles of programme management and the view that each programme's management is unique and that each situation must be analysed separately.

REFERENCES

Flyvbjerg B (2014) What you should know about megaprojects and why: an overview. *Project Management Journal* **45(2)**:6–19.

Hanford M (2004) Program management: Different from project management. See http://www.ibm.com/developerworks/rational/library/4751.html for further details (accessed 07/10/2014).

Programme Management in Construction
ISBN 978-0-7277-6014-2

ICE Publishing: All rights reserved
http://dx.doi.org/10.1680/pmic.60142.049

Chapter 4
Case studies

4.1.   Description

This chapter will describe and outline several case studies of actual development and construction programmes. The programme in which the first case study is reviewed has been ongoing over a period of 46 years in the USA, the third case study programme ended with the completion of a landmark resort in Dubai, UAE, and the others were still under construction at the time of writing.

As discussed in earlier chapters, a programme is a group of related projects managed in a coordinated manner to obtain benefits and control not available from managing them individually. Programmes may include elements of related work outside the scope of the discrete projects in the programme. Certain projects within a programme should deliver useful incremental benefits to the organisation before the programme itself has been completed. Programme management also emphasises the coordinating and prioritising of resources across projects, managing links between the projects, and the overall costs and risks of the programmes. Programme management may provide a layer above the management of projects and it focuses on selecting the best group of projects to implement, defining them in terms of their objectives and providing an environment where projects can be run successfully. Programme managers should not micromanage, but should leave project management to the project managers. Hence, this chapter will focus on some programmes experienced by the authors and look at the successes and failures of those programmes.

Over 30 years ago in the development and construction industry, the general way of describing a programme was to refer to it as a 'project'; no matter the length of time involved. 'Programmes' were, however, paramount in different industries, such as those NASA was producing in the 1960s, 1970s, 1980s and 1990s – the Gemini Programme, Apollo Programme and Space Shuttle Programme to name a few. All of these were referred to as programmes because multiple launches or projects happened within those programme lifecycles. Although the programmes had no firm ending dates, the specific launches did. One of the defining features of a programme is that it has no firm ending date, whereas projects do have firm ending dates.

Many parallels can be drawn between the space programmes and construction programmes. The space programmes were conglomerates of many entities, formed to produce interim results by means of a management structure that set for each launch a milestone date within an open-ended long-term programme. The management structure for each programme was designed to achieve its objectives: place a person in outer space, then land on the moon; finally re-use the spacecraft by flying back to earth as the Shuttle Programme completed. The programmes were stopped when they became obsolete or when the final goal was realised.

A development or construction programme exists to satisfy the goal of the client. It does that by producing outputs called projects. A project is always defined by scope, schedule and cost. The

Figure 4.1 Control responsibilities

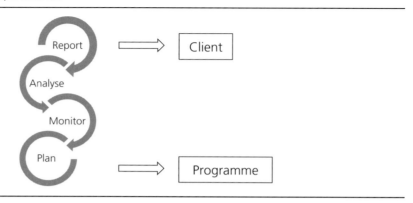

scope and schedule always affect the cost, but the schedule always affects the ability to complete the original scope. Typically, during financial feasibilities the start and finish timelines are considered sacred for any project due to the added cost of borrowing money for longer periods of time.

Programmes only have a goal or objective as the ending, considered open with regard to time, but with a focus on delivering a positive outcome in the company's performance. A programme should be thought of as the global control of multiple similar projects where the programme manager's responsibility is to plan, monitor, analyse and report to the client on all projects in an effort to allow the client to affect informed decision making with respect to the direction of the programme, as shown in Figure 4.1.

The key difference between a project and a programme is the ending of the timeline. The project has a finite ending date, whereas a programme continues without a preordained ending. The ending of the programme is usually set with the client upon his or her decision to end the programme. Many reasons can influence that decision, such as there being no more projects to produce, the attainment of all goals, significant changes in the initially targeted outcome due to government restrictions, loss of funding, or stakeholder changes. Typically, the programme completion is based on a decision rather than a pre-determined date, as with a project.

The standard way to explain the difference between programme and project management can be summarised thus: 'project managers manage projects and programme managers manage a portfolio of projects'. Such definitions leave much to be desired. They have led to the myth that programme management is glorified project management.

In fact, there are five fundamental differences between a programme and a project:

- **Programmes are ongoing; projects end.** Programmes usually span a far greater duration than do projects. This might seem like an arbitrary difference. However, managing a programme involves long-term strategic planning that is not required of a project. The ongoing nature of programmes also means that they engage in continuous process improvement and change.

- **Programmes are tied to the organisation's financial calendar.** Programme managers are often responsible for delivering results tied to the organisation's financial calendar. Projects run on project time. A project manager is not responsible for delivering quarterly results, merely to deliver the project on time and within set budgets. Programme managers are often driven by quarterly and year-end results, as are managers in the rest of the business.
- **Programme management is governance-intensive.** A programme typically is governed by a senior-level board that provides direction, oversight and control. Programme managers must be able to influence at this level. They must also facilitate resolution of disagreements between executives. The programme manager must ensure that the governance board provides achievable objectives for the programme, and must deliver to these objectives. Projects may have a similar governance structure. However, they tend to be less governance-intensive.
- **Programmes have greater scope of financial management.** Projects typically have a straightforward budget. Project financial management is focused on spending to budget. Programme managers may be responsible for revenue and costs that are critical to an organisation's financial results. Budget planning, management and control are significantly more complex in the context of a programme.
- **Programme change management is an executive leadership capability.** Projects employ a formal change management process. Programme change is more difficult to manage due to its typical impact on the overall strategies of the organisation. Programmes are driven by an organisation's strategy. They are subject to market conditions, government regulations and changing business goals. At the programme level, change management requires executive leadership skills.

There are many opinions and views on the true definition of programme management as it relates to project management. Overall, most agree that programmes are the systematic global management of a series of projects that are selected to promote a positive outcome for an organisation. Many elements of both are managed in the same way, although a programme is conceived to accomplish an organisation's mission such as sustained profits or social benefits.

Projects have a locked scope and are constructed with time being of the essence with regard to completion. Changes are usually minor and are implemented with pre-agreed procedures within the contractual arrangement. Programmes, however, may change scope and acceleration as outside conditions, such as economical and business environments, warrant.

Other key factors determine a programme rather than a project; an important indicator is that programmes are set up to continuously produce output based on the client's vision, such as the production of similar products over a long period of time. In the construction industry, these results are usually the completion of similar industrial, infrastructure or building projects. This can be easily seen when looking at what land and real estate developers do to produce revenue. In contrast to the space industry, the construction industry's typical goal is to produce finished infrastructure and/or building structures for the specific purpose of turning profits (positive outcomes) for the company.

If one examines the typical business model or government organisation of major and mega construction developments one may see that setting up a programme to manage the completion and long-term life of the enterprise is the only way to achieve success in an efficient manner. With that thought in mind, case studies displaying typical scenarios that a programme manager might encounter in the business of developing, constructing and running mega programmes in the construction industry are presented in the remainder of this chapter.

Large development companies and government entities, which are targeting multiple additions to their portfolios, are obvious candidates and should consider structuring their management organisation as programme management to affectively control their resources. Portfolio management is, generally speaking, similar to programme management, although the two do not require the same level of management; portfolio management can be described as the management of capital building projects from inception to delivery, then through facilities operations. The word 'portfolio' refers here to a company's multiple property assets, which are under continuous operation for the purpose of generating positive outcomes in the form of profit.

The case studies revolve around four actual programmes for major corporate and government development and construction programmes. The authors will outline and explain why they consider them programme management, not merely large projects or portfolio management. The various programmes are related to creating tangible assets for a wide variety of entities: government housing authorities, government school authorities, residential and commercial condominium developers and resort hotel developers. These case studies were chosen due to their specific management structure and/or frameworks that were unique in each instance. Each case study has a different outcome planned, and demonstrates that there are many ways and means to accomplish a positive result. No cast-in-stone formula is the right way; more often the framework of the programme will be accomplished by setting up what works within the confines of that particular organisation (Flyvbjerg, 2006). We will explain in detail each case study along with the programme outcome results.

4.2. Case study: Programme One
4.2.1 Narrative
This first programme case study was conceived by a real estate developer with the idea of producing a landmark luxury residential condominium property that would create positive outcomes (profits) for his company by the sale, construction and closure of continuous outputs or buildings.

The developer formed a joint partnership with a financial trust and a large real estate company based in New Jersey, USA. The partnership confirmed him as managing partner of the programme and assigned him the task of planning, designing, building and selling the partnership's products, in this instance the condominium units. He was responsible for the mitigation of financial, schedule and technical risk, as well as for the standardisation of design and concept, setting strategy, tasks and milestones, procurement of vendors and contractors, managing the ongoing portfolios and communicating the performance to his shareholders. In other words, he was the programme manager.

The land was already selected before the due diligence was started: a piece of waterlocked property in the middle of Miami, Florida, USA. This property constituted an island which had already been partially developed by the previous owners and two adjacent parcel owners. The net area that could be developed was approximately 81 hectare. At the completion of the land planning, the land was divided into many parcels distributed along the perimeter of the island, which created spectacular water views for the majority of the units and provided the central areas with an open green belt for use in the future as a park or executive golf course. The amenities for the luxury condominium venture included hotel rooms, commercial retail centres, an executive club, parks, tournament-level tennis centre, deep-water yacht marina and five-star restaurants to support the residents. The net yield of the land plan was 1200 units, each being approximately 743 m^2 in size and spread over multiple parcels and buildings.

Transportation infrastructure was a significant issue due to the unique nature of the access for the property. No bridges had ever been planned because this had originally been a private island with no public access. That in itself made this one of the most secure properties in the world. One of the ultra-high-end selling points of the condominiums was the private transportation network of ferries for vehicle traffic and 12-m yachts for foot traffic. This private transportation network ensured a high level of security for the residences.

This programme is a good example of why certain real estate projects that have an ongoing development process are really better described as development programmes due to the open-ended completion and continuous product output. In the era in which this programme started, construction managers and developers explained development programmes like these as phased projects. Today, we better understand them as programmes because the timeline is more a factor of market conditions than a cast-in-stone completion date. For example, the release to construct condominium buildings is governed by the sales of units. Typically, financial institutions will not lend money against unsold building units at one time with the hope that sales will keep up, as this is too risky. Therefore, even after final planning has defined the number of buildings, they will not commence money-lending until market conditions dictate success of the financial outcome.

Originally, this programme had been thought of as a viable venture in 1968; however, due to a high-level national political office held by one of the property owners at that time, the due diligence period was delayed for five years, until that certain owner was less visible to the public. This development programme can be described as a planned residential condominium and commercial venture in which land use planning finally started in 1973. The programme's initial delay of five years was due to outside political influence. The original feasibility study planned the complete build-out of all 1200 units to be five years from the date of the final approved land plan. Again, the commencement of the physical construction was delayed due to the length of time for accomplishment of the final land plan.

Planning for the concept took seven years to complete due to failed government approvals for zoning and density. The first land planner involved never produced a viable design that would net the necessary saleable space. Some designs were so unusual that the government agencies were hesitant to return answers quickly. One such design resembled Italy's famous city of water, Venice. This was a critical element for the programme to track due to its serious time delay impacts. When the final land planning was completed in 1979, 11 years had passed from the initial concept. The market conditions had changed significantly. A new president was in the White House and interest rates were at an all-time high, at over 20%. Fortunately, the return on investment originally calculated was still valid due to the high margin potential of the condominium units.

Today, the property build-out still remains uncompleted due to many outside influences over the years such as down-market conditions, liquidation of the original financial backer, lawsuits and foreclosure actions from the civil court system. The original development company was dissolved in 2004 following the financial stakeholder's liquidation of assets and resold to another organisation with the same goals. At the time of writing, in 2014, the remaining parcels are free from previous encumbrances and plans to finish the final building are being sorted out with the new owners. The current number of units has been modified to be limited to no more than 900 compared to the original land plan of 1200 (Figure 4.2).

4.2.2 Data

The original land plan consisted of the following facilities:

- condominiums: 1200×743 m^2 luxury units spread over multiple parcels, designed using only mid-rise buildings
- hotel: 200-key five-star
- club: the existing mansion and compound to be transformed into a five-star resort-type club facility
- commercial space: approximately 9290 m^2 retail space and commercial space
- marina: 100-slip deep-water yacht basin
- sports centre: world-class 17-court grass and clay tennis facility
- support facilities: maintenance support for all groundskeeping and common infrastructure, vehicle and pedestrian marine transportation and building maintenance and fire station.

4.2.3 Management structure

The management structure at its inception was typical of those characterising similar develop-ments, with the programme manager's role being defined as the developer and/or general partner of the development company; this typical development organisation is shown in Figure 4.2. This was a time well before the construction industry understood the concept of programme manage-ment. Today, this general partner would be the programme manager. The general partner or programme manager is the individual responsible for planning, monitoring and analysing the programme's performance and reporting progress on goals and objective to the partners. This includes all aspects of the business enterprise such as design, construction, government relation-ships, sales, marketing and operations.

4.2.4 Staff

Development manager: his role was to manage all aspects of the physical creation of infrastructure and buildings: scope, schedule and budget. He was the central point of contact for everyone from the contractors to the project managers. The development manager's responsibilities included directing the design consultants for design direction handed down from the general partner, conveyance of systems and compliance of regulations with the various government agencies, monitoring the contractors on progress, and reporting performance to the general partner (see Figure 4.3).

Deputy manager: his role was to coordinate between the contractors, designers and agencies to ensure satisfaction of design intent, compliance with regulations and building codes.

Project manager: his role was to monitor and report on the contractors' progress with regard to scope, schedule and budget for specific projects.

4.2.5 Milestone dates for condominiums

The original time durations in the feasibility study indicated build-out within 60 months of final zoning approval. From the actual timeline shown in Table 4.1 it is apparent that the initial plan was way off the mark. Prior to actual real estate title transfer of residence and permanent occupancy on the development, the infrastructure and transportation systems had to be in place

Figure 4.2 Management structure of condominium developer

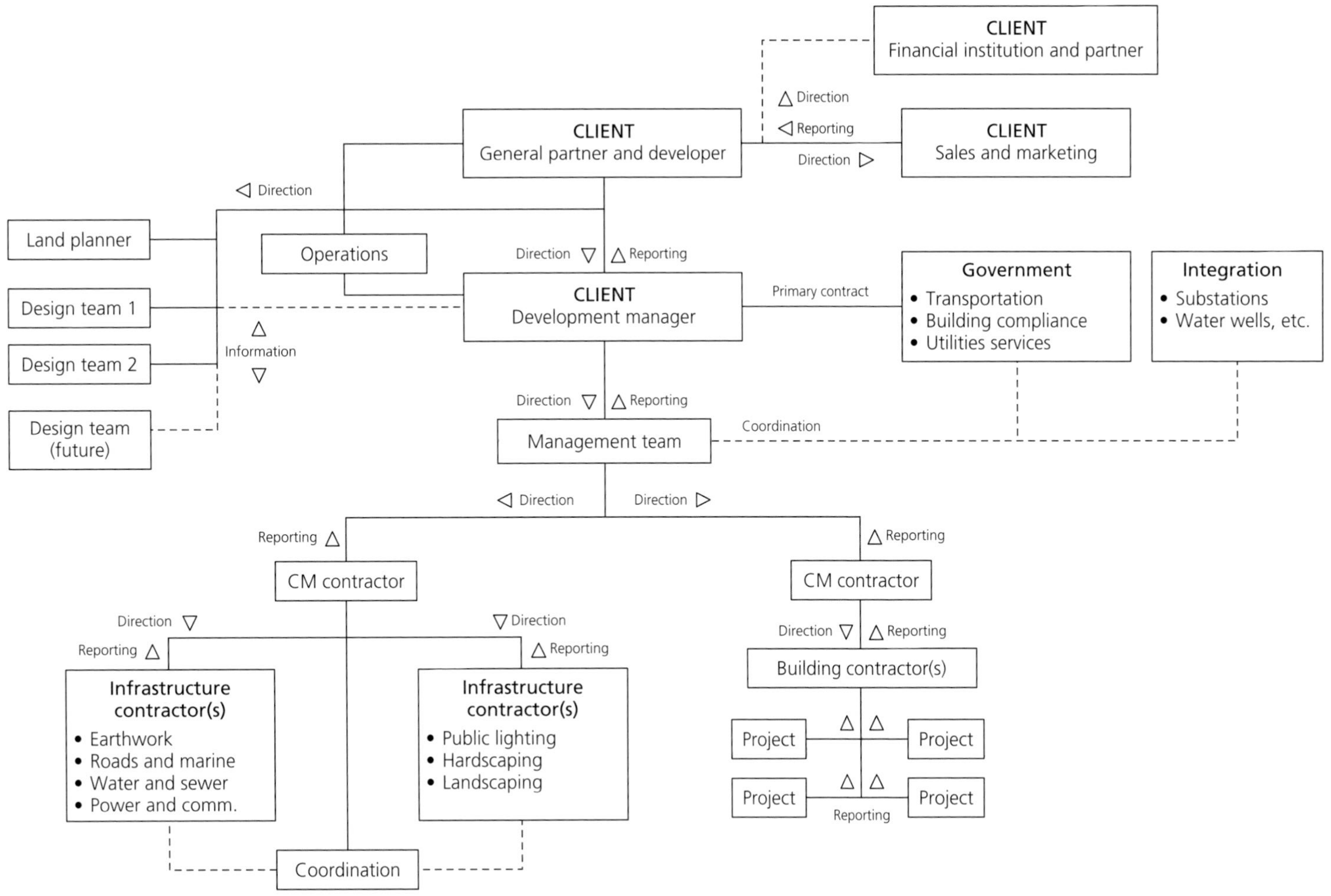

Table 4.1 Milestone dates for condominium development

Activity Description	1968	1973	1979	1980	1981	1982	1983	1984	1985	1986
Concept	×									
Land use planning		O	—×							
Land planning and zoning approval			×							
Land purchase			×							
Development company opened			×							
Construction of transportation facilities					O	——×				
Construction of roads and utilities						O	——×			
Infrastructure and utilities services conveyed								×		
Access systems, car ferries, barge landings commissioned									×	
Completion of amenities, club, marina										×
Completion of first condominiums										×

and operational to support the residence. This would make up the initial milestone activities, which are listed above.

4.2.6 Pricing and cost control

The original purchase of the property was US$13 200 000, compared to the commercial Appraisal Institute (MAI) appraisal of US$100 000 000. This allowed the organisation to continue the development processes even though high interest rates had slowed down the financial institutes in lending money. This was critical due to the large amount of infrastructure needed to design and build prior to any real estate closings (revenue).

4.2.7 Actual cost

At the time in which the original development company ceased operations and shut down capital contributions, what was shown on state records amounted to US$463 421 491 invested. Floor plans were added over the years to allow a wider range of saleable unit plans. Over 560 units had been sold at that point and selling prices ranged from US$600 000 to US$6.4 million. The overall net result of the programme created an outcome for the development company four times that of the investment.

4.2.8 What went wrong and right

One obvious flaw was the disproportionate length of the planning phase. This was due to the land planner's poor concepts and heavy regulations from multiple authorities and jurisdictions complicated by some having overlapping powers. The planning took seven years to complete. This delayed the physical development phase, which created unpredictable interest rates and mounted unplanned debt without return on investments. This unfortunately was not due to the structure of management, but to the red tape of government agencies and the land planner's ill-conceived ideas.

In a telephone interview, the original developer stated that the final land plan was only completed when he realised the planner was waiting too long for the process of submitting ideas and waiting on answers from the authorities. At that point, he formulated a workable plan with his architect and then worked closely with the authorities to receive approval. His advice to me was that success is formulated by finding a workable solution to problems. Do not rely on traditional processes to navigate through roadblocks obstructing the programme's movement forward.

Another issue that created irreparable damage to the overall programme came up during the planning and procurement of the various landing points for the marine transportation systems; a certain parcel of land on the mainland was available that would have been turned into parking for island residents and visitors. The general partner or programme manager requested from the financial institution a minimum of US$6 million to secure the purchase of the necessary land. The request for additional funds by the general partner were viewed as overstated by the financial manager. Subsequently, the financial manager did not agree with the amount requested and only made US$3 million available. This resulted in the loss of the property to other parties for US$4 million. The management breakdown was due to the fact that the financial institution assigned only one individual to act as the governance for the release of necessary funds. Overall, this was a highly successful programme due to the focus of the original developer on delivering the right product to the market despite massive amounts of bureaucratic red tape and changing market conditions. The organisation would have benefited from additional middle management and a strong steering committee but this programme had a leader with all the necessary attributes a solid programme manager should have: financial understanding of the market associated with the programme, cross-cultural awareness, leadership, communication skills, influence, negotiation, conflict resolution skills.

Moreover, this programme manager had a vested interest in the success of the programme, which instilled in him the drive and fortitude to push the programme forward. It is imperative that the programme manager should believe in only one agenda, that of the success of the programme.

4.2.9   The benefits of programme management application
Whether he or she be called the developer or the programme manager the roles are the same: managing a collection of resources – money, materials and labour – to affect a positive outcome for an organisation's financial or social goals. The organisation structure depicted in Figures 4.2 and 4.3 is lean compared to what is thought of today as a proper programme structure. Missing components from the management structure perspective include, in the upper hierarchy, the oversight board or steering committee. This programme would have benefited greatly by appointing a board of trustees to manage the financial needs of the development, thus minimising the short-sighted decisions of individuals that created negative impacts.

Furthermore, the senior and middle management ranks in the typical real estate developer organisations are most often reliant on single individuals having too much responsibility for tasks that directly impact outcome. This lack of governances without the benefit of checks and balances can create situations that spiral on, when input from other sources may put the programme back on course. The modern ideal structure of the programme management framework allows for great transparency throughout the organisation, consequently building in the checks and balances through the process of installing certain governances.

4.3.   Case study: Programme Two
4.3.1   Narrative
This case study should be considered a programme due to the reason it was conceived: to create a positive effect with regard to a developing country's long-term goals for the younger sector of its

Figure 4.3 Organisational structure and workflow responsibilities

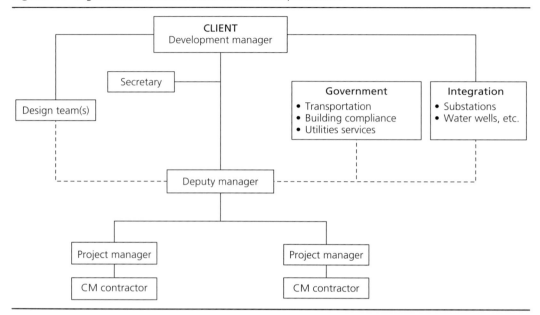

population by providing them with a high level of education. The government's aims were to cultivate a highly skilled and motivated potential workforce. In other words, the programme's aim, as stated by the government, was to bring about a positive outcome for the country by producing schools or outputs for social benefits.

This country's educational system is currently going through an important phase in its history due to a design and construction programme that been decreed by the government. The government, acting as the programme manager, will plan, develop and construct up to 4000 schools throughout this vast country in an effort to strengthen the country's future workforce. This programme is called the Project for General Education. Although called a project, it is actually a programme due to the open-ended completion time, long-term production of multiple similar buildings (output) and long-term operational character (social benefit). No actual date of completion has been set, but an estimated six years was projected for completion.

All sites consist of one, two or three similar 5 km² main school buildings with more than 20 modern classrooms, science laboratories, computer laboratories, language laboratories, home economics, library, offices for supervisors and staff, and support areas. An 850-m² multi-function building, canteen and playground will complete the site, surrounded by a perimeter wall and security gate.

A large number of schools in three separate provinces have been selected for the implementation of the first series of projects within the massive multi-billion-dollar programme. The first projects in the programme consist of 2000 schools and will be implemented over six years with no specific timelines, only the specification of a 14-month construction period. Therefore, it is correct to say that those are projects, due to their defined timeline, scope and budget. As pointed out in other chapters, however, these might also be viewed as constituting a programme, due to the layer of management above the project managers who are coordinating and prioritising resources and acting as the link between all individual projects with respect to the risks of budget, time, technical issues and impact on the overall programme outcome.

4.3.2   Data

The full programme is estimated at around 4000 schools of which over half have not been integrated into the programme at this time. The amount has only been estimated because the full extent of available resources is not currently known. Selections of sites will ultimately be made based on need and priority of the region. Many small releases of projects have been assigned to multiple groups of consultants and contractors at a rate of two to four schools each. This has created many problems for the client, who must analyse and disseminate information from a variety of sources comparing actual progress with the planned schedule.

In an effort better to control the progress, large groups of sites have been released for development at a time. The first group comprises 400 school projects on 280 sites, the facts of which are listed in Table 4.2. This first group was assigned to one supervision consultant and one contractor. The client had the understanding that management of the multiple construction sites would provide a more transparent and efficient information flow, allowing the client to analyse progress in a more accurate manner. Additionally, efficiencies in procurement and shared management should be beneficial.

The 400 schools of the programme are mainly concentrated within the central region. The remaining balance is in the two coastal provinces. Release of sites to the contractor was dependent on progress. Poor progress necessitated the reallocation of certain sites to different areas. The entire programme spans hundreds of thousands of km^2 of the country. The education department within the government are in charge of the programme. Their current plan is to move some of the coastal region projects to the central region due to delays created by the contractor by his lack of organisation and unsatisfactory resource management.

4.3.3   Structure

The management structure of the 400 schools programme may resemble that of a large project; however, it has many cross-dependencies such as design and procurement. The organisational chart for the consultant is shown in Figure 4.4.

Table 4.2 List of schools by group – the 400 schools are sub-divided into groups within regions as listed below

Description	Sites	Schools
Group 1 (central)	32	57
Group 2 (central)	31	67
Group 3 (central)	50	59
Group 4 (central)	44	64
Group 5 (central)	26	32
Group 6 (central)	31	42
Group 7 (coastal)	26	33
Group 8 (coastal)	22	26
Group 9 (coastal)	18	20
Totals	280	400

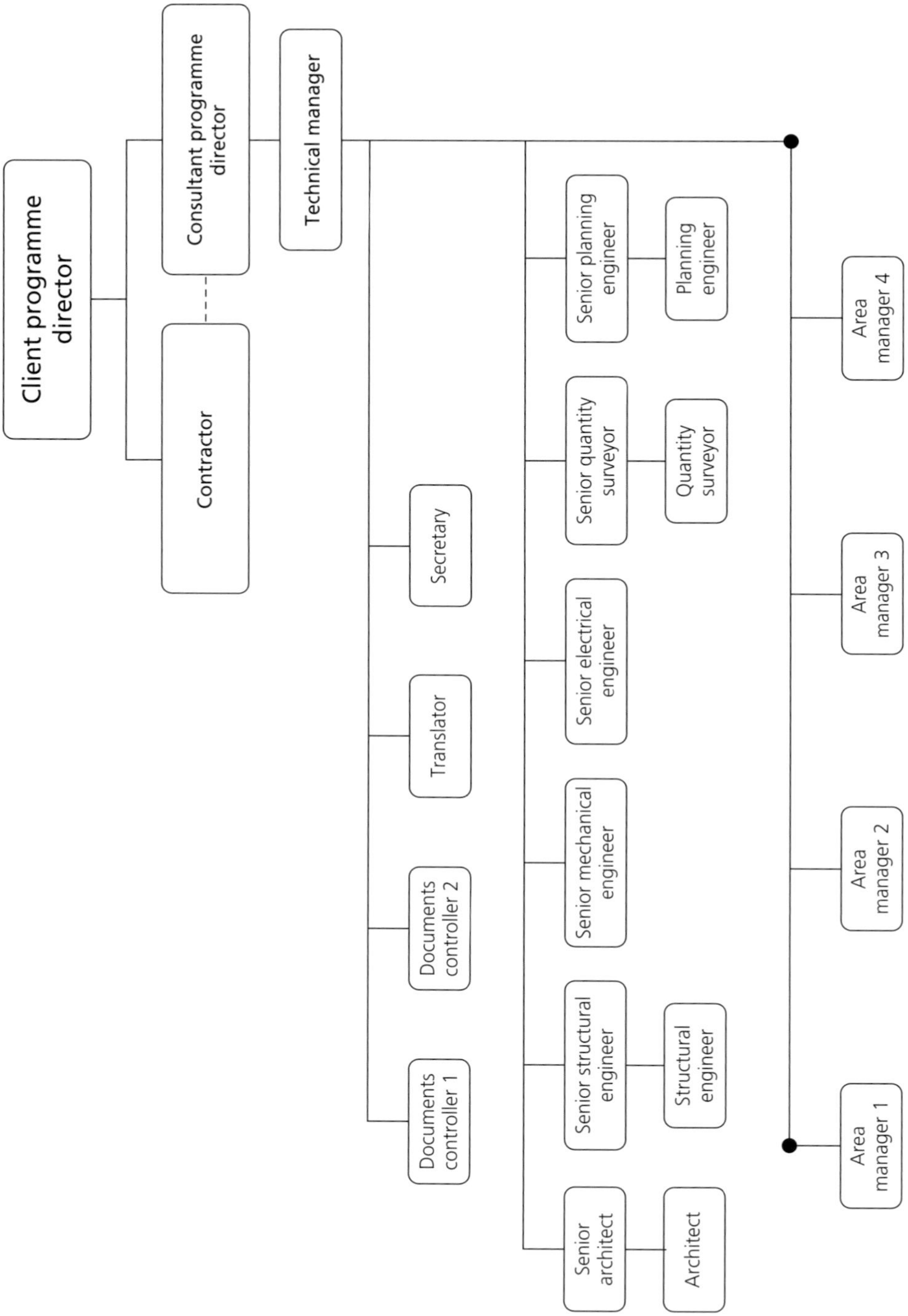

Figure 4.4 Consultant organisational structure

Terms used in this case study are as follows.

Department of Education: has taken on the role of owner and client, and that of the authority acting to exercise the regulatory and proprietary authority on behalf of the country. This includes all facilities located on the building site in accordance with the laws, ordnances and codes having jurisdiction.

Consultant: described as the 'authority's representative' which means the owners' representative for the Department of Education in all business concerning the design, planning, construction and delivery of all 400 school facilities. The main task is to monitor, guide and steer the construction process. The main responsibilities shall be to ensure quality, provide technical advice, review and approve construction documents, maintain budget control and on-time scheduling.

Design consultants: are the architects, engineers and designers responsible for completion of all design and construction documents necessary to convey the magnitude and intent of the scope of work to the owner/client and contractor. This will include conceptual drawings, design development and construction drawings.

Contractor: is the prime general contractor responsible for completion of the entire scope of work within the scheduled timeline set by contract. The general contractor is obligated to provide all resources necessary to meet the schedule and supervise all services in connection with the construction of the project.

The strategic management plan is to provide the client with project management coverage for the design completion and construction implementation period. This will be accomplished by a core team of engineers, architects, planners, quantity surveyors, document controllers and a translator acting as a direct link to the contractor's senior management. The core team is coordinated by a full-time technical manager, who reports directly to the supervision consultant's programme or project director. The core team provides the client's programme or projects director and his staff with senior-level management assistance. The consultant provides on-site engineering coverage for each site directly linked to the core team.

The consultant's plan was to link the entire programme staff via an internet-based information management system to all project offices, core team and the client. Project sites would have been able to review document logs, approved material data, specifications, shop drawings and schedules through the web-based information system as needed. Unfortunately, the way in which the client contracted certain obligations such as the information system created a situation where the system was never purchased. This left the programme with multiple segregated systems with no programme-wide view of tasks and progress, poor availability of information and an incomplete audit trail.

4.3.4   Staff
The core team consisted of 13 full-time senior staff consultants working closely with the client and design consultants stationed in the client's office complex with the contractors' core staff. The core team has been tasked with managing, interfacing and supporting the site groups for inspections, quality, safety, scheduling, shop-drawing approvals, document control, payment/variation requests and general construction issues. The consultants' programme or project director has been tasked with interfacing with the client's programme or project director to assist in the overall

construction management of the project. The core team has been tasked with the following:

- construction contract compliance
- supervision of site groups
- technical support
- compliance with programme specifications and drawings
- review or approval of shop drawings
- review or approval of material submittals
- review or approval of monthly interim payment applications (IPA) and contractors' request for variation orders (CRVO)
- document control
- advice and consultation to the client for technical, risk and management issues
- progress status communication with the client through weekly and monthly reporting
- providing the client with solutions to resolve problematic issues.

The consultants' site supervision management structure consists of 280 engineers overseeing the direct work of the contractor, organised by group with one area manager heading the engineering staff of all schools within that group. As each new group is released for construction commencement an area manager and staff will be assigned. Each site engineer is stationed on site in an office supplied by the contractor. Each group area manager is stationed in a regional office with his supervising engineers and staff engineers. Direct supervision of project sites is carried out by the area managers, which shall report to the technical manager and subsequently to the project director. The area manager is supported by either two or three supervising engineers, which shall handle up to eight schools each. Three or four electrical and mechanical engineers rotate throughout the group. Permanently based on each site is one site engineer. The site team will be specifically tasked with the following:

- construction supervision of project sites
- compliance with contract through monitoring, inspecting and reporting
- verification of quantities through measurement
- threshold inspection or approval
- witness and review of test and laboratory results
- progress monitoring/or reporting
- quality assurance through monitoring, inspecting and reporting
- safety compliance through monitoring, inspecting and reporting
- commissioning and turnover acceptance and certificate.

4.3.5 Milestone dates for schools

The contract for construction with the general contractor considered the commencement of the timeline clock for each school to begin with the client handing over the survey, soil report and necessary handover documents. The agreed timeline for each site was 14 months. The client decided to release the projects by groups starting with Group 1; this would allow a more structured release, staggering the commencements and therefore resulting in staggered completions and handovers (see Table 4.3).

4.3.6 Pricing and cost control

The construction contract documents consisted of a governmental standard form contract similar to a Fédération Internationale Des Ingénieurs-Conseils (FIDIC) layout with sub-documents

Table 4.3 Milestone dates for school groups 1–8

Activity description	2009	2010	2011	2012	2013	2014	2015
Land purchases		×					
Design of prototype building	×						
Tender of general construction contract		×					
Contract for project management		×					
Commencements of groups 1–4							
Completions of groups 1–4				×			
Commencements of groups 5–8			×	×			
Completions of groups 5–8					×		

attached such as a bill of quantities, standard client specifications and a set of incomplete detailed design drawings. This incomplete package created confusion and conflicts between the contractor and the client throughout the material selection process, which continued into the implementation period.

The bill of quantities was prepared by the client without input from the contractor prior to the tender with the contractor. The specification was a standard form using generic description for much of the finishes. The detailed design drawings were not completed due to the original design architect being discharged from the programme before the packages became complete.

Although all of these problematic issues were present, the contractor and the client both agreed to commit to the contact with the understanding that the specific details would be resolved as the programme moved forward. The partial bill of quantities is summarised in Table 4.4. The currency shown is US$.

4.3.7 Actual cost
As outlined above, the construction contract documents were incomplete from the inception of the contract agreement. The contract provided for the contractor to notify the client within 10 days of discovering an item which was not included or indicated on the drawings, bill of quantities or specifications. In light of the fact that the total contract package was less than complete, the contractor discovered many discrepancies throughout the construction process that were not properly covered. In doing so, the contractor triggered a process prior to approval by the client for clarifying and pricing any and all changes to the subject item.

Poor organisation and management by the contractor for cost control change covered up the actual magnitude of the budget escalation. By the conclusion of the contract time period, the amount of the changes, according to the contractor, had escalated to 16% of the cost of the work in place. This escalation of the budget was not anticipated due to the fact that the contract only provided for a 10% increase, whereas a claim for the balance of 6% is now anticipated.

4.3.8 What went wrong and right
The programme was plagued with poor planning from the beginning of the formation of the construction contract documents. During the pre-planning stage, finalised budgets had not been released with the finance department, the final interior designs and specifications had not been approved and various land parcels had not been secured. Consequently, the commencement of the

Table 4.4 Summarised Bill of Quantities for Groups 1–4

Group 1			Group 2			Group 3			Group 4		
Site	Building count	Contract amount	Site	Building count	Contract amount	Site	Building count	Contract amount	Site	Building count	Contract amount
1	1	6 983 637	1	1	6 114 224	1	1	7 020 467	1	1	6 138 819
2	2	12 475 345	2	1	6 131 871	2	1	7 363 014	2	1	7 013 854
3	2	12 344 045	3	1	6 281 305	3	1	7 390 975	3	1	7 071 573
4	1	7 608 791	4	1	6 774 536	4	1	7 394 275	4	1	7 290 193
5	1	6 363 205	5	2	11 808 095	5	1	7 400 612	5	1	7 335 588
6	2	12 470 869	6	2	11 821 900	6	1	7 403 353	6	1	7 426 563
7	3	18 691 552	7	2	11 857 052	7	1	7 458 872	7	1	7 501 014
8	2	13 501 072	8	2	12 086 073	8	1	7 491 666	8	1	7 569 995
9	2	12 411 501	9	2	12 497 948	9	1	7 568 706	9	1	7 701 368
10	1	6 251 498	10	3	17 533 033	10	1	7 589 093	10	1	7 715 843
11	2	12 539 897	11	3	17 823 639	11	1	7 597 859	11	1	7 718 079
12	2	12 259 426	12	3	17 835 712	12	1	7 653 910	12	1	7 761 556
13	1	6 200 770	13	3	18 013 740	13	1	7 667 529	13	1	7 906 277
14	1	6 424 220	14	3	18 737 836	14	1	7 757 643	14	1	8 037 823
15	3	18 317 231	15	4	23 228 908	15	1	7 757 822	15	2	13 121 109
16	2	12 599 202				16	1	7 784 560	16	2	13 352 466
						17	1	7 987 785	17	2	13 796 186
						18	1	8 028 125	18	2	14 041 025
						19	1	8 042 384	19	2	14 527 338
						20	1	8 125 987	20	2	14 598 278
						21	1	8 144 168	21	2	15 017 956
						22	2	12 780 582	22	2	15 050 505
						23	2	12 962 014			
						24	2	13 050 240			
						25	2	13 574 801			
	28	$177 442 261		33	$198 545 872		29	$212 996 442		30	$217 693 408

general contractor's work was delayed and subsequent work flow was interrupted due to the lack of an agreed master plan of action agreed by the client and the general contractor. The stated objective of the contractor was to take over all 280 sites within the first eight months of the construction schedule. Only 157 sites were released by the client, however, due to the contractor's lack of progress and to certain land parcels not being ready to release for construction.

Subsequently, the general contractor proved to be unqualified to carry out the work properly due to his poor management and lack of resources. Their management structuring indicates that the contractor's core team would have direct control over the field construction work through their regional offices. Structuring the management in this configuration, from the top down, was believed to be the best way of controlling the work. The structuring of the management failed to be carried out, however, as the contractor's director maintained direct control of the field, rendering the core team's management as ineffective.

The contractor's core team manager was not given clear empowerment to manage the field operations, which resulted in the field manager pushing the work improperly without following proper protocol and procedures. This resulted in a large number of non-compliance reports (NCR) due to low-quality workmanship. The contractor's programme director and project managers made several promises and commitments to correct the management deficiencies concerning all sites. These deficiencies are as follows:

- Hire an independent consultant to test, create methods statements, and certify structural deficiencies outlined in open NCRs.
- Build regional offices.
- Pick up the pace of construction and deploy additional labour.
- Submit proper material to match design intent.
- Submit proper shop drawings in a timely manner.
- Implement a proper quality assurance and quality control for the programme.
- Distribute shop drawings to all sites.

4.3.9 The benefits of programme management application

It should be obvious from the information described above that this project suffered from a myriad of problematic issues. This created a multitude of delays on both the client's side and on the contractor's side. The client and all stakeholders would have benefited greatly from a well-organised programme management approach that made sure all basic fundamental elements were in place prior to the commencement of the programme.

The missing fundamentals were several key programme elements, which would have resulted in better-quality information from the field level being analysed and reported to the top stakeholders, thus allowing the timely making of informed decisions and then subsequent passing on of transparent direction.

These elements and resulting failures were as follows:

- issue: lack of programme execution plan
    - failure: risk to schedule, financial and technical goals
- issue: uncompleted construction documents
    - failure: risk to scope, budget, schedule and procurement

- issue: missing global documents control system
  - failure: communication and conformity.

In the client's rush to produce a fast outcome, he ignored some of the basic fundamentals of both project and programme management. Implementing a programme execution plan (PEP) would have created a well-defined road map for all stakeholders to follow, and provided a predetermined matrix of responsibilities and obligations. Completion of the construction documents prior to commencement of work on any site would have minimised cost escalation and procurement mistakes by the contractor. In today's fast-paced environment, the management of a programme is less inefficient without an integrated collaborative platform and control system to tie together all stakeholders. Implementing a computer-driven system from day one to manage information, process and make documents available to all necessary stakeholders is without a doubt a mandatory element of a proper programme. Lower productivity, increased errors and exposure to higher risk will result in the absence of the single integrated platform.

4.4. Case study: Programme Three
4.4.1 Narrative
This programme was conceived by the chairman of an existing and highly successful hotel company that was interested in starting a new brand of themed properties, which would be recognised around the world. This enterprise, thus, matched one of the defining elements of a programme: creation of outputs to result in a significant positive company outcome.

A large island in the Bahamas with an airport and airline, multiple hotels, golf course, residential and commercial properties was purchased from the Resolution Trust Corporation (RTC) by an international developer and hotel operator. The chairman through the company's organisation purchased over 242 hectares of this island in order to develop a new iconic destination resort and retail or commercial tracks. The initial purchase in 1993 included several functioning hotels and the existing operation of a large destination resort, which had long been in need of capital improvements. The purchase included the water desalination plant, sewage plant, golf course, airport and the road systems for the entire island. All infrastructures were in place, although not up to the standard required for the level of service envisioned for the future.

Over the previous years of operation, before the sale, all properties had been operated on a tight budget leaving everything in need of repairs, replacements and modernisation. The challenge here was to turn the main resort into a modern themed hotel, which would serve as the model for the company's product line and a source of financial resources to facilitate the future redevelopment of the island and the future development of other properties the company was planning to purchase for their global expansion. The chairman of the company convinced his private investors to fund the renovation of the main resort due to his highly successful track record and influence in this type of business.

The first phase was staged in three parts, all done simultaneously; first, to complete renovation and new construction of the existing three-star high-rise towers, which had 1200 rooms or suites, add major landscaped water gardens with interactive rides, renovate the existing 2790-m^2 casino, create abundantly stocked 13.25 million litre exotic fish saltwater display lagoons and renovate the existing convention centre. The second phase would involve the completion of the renovation of a famous five-star 100-room boutique hotel with villas, restaurant and recreation facilities. The third phase would involve completion of the long-needed maintenance and updating for the reverse osmosis water plant and sewage treatment plant.

This was an established hotel company with departments to manage the various functions such as hotel operations, finance, procurement, marketing and development. The due diligence for the entire island property was handled by the development department with assistance from both operations and procurement departments prior to the sale offer. Upon real estate closing, the hotel operations had formulated a plan and were ready to move in and assume control.

The plan was a monumental undertaking due to the timeline and operational requirements of the hotel company. The launch of the new brand was scheduled for December 1994 in time for the start of the annual holiday season. The hotel was to remain in operation during the renovation in order to maintain a presence and mitigate revenue loss due to the renovation.

4.4.2 Data
The purchase of the properties included the following assets:

- three-star resort   1200-room hotel and casino
- five-star hotel   100 rooms
- three-star hotel   150 rooms
- golf course   18 holes
- airport   one private commercial runway
- airline   multiple aircraft designed to land on a runway of a short distance
- utilities plant   reverse osmosis plant and sewage treatment with total island network
- road system   complete road network throughout the island
- restaurants and clubs   multiple properties throughout the island
- miscellaneous   raw land, residences and apartments throughout the island.

4.4.3 Structure
The management structure was set up as a property development department through the hotel company. An executive vice president of the company was appointed to manage that department with direct input from the chairman, who was acting as the visionary and steering board leader. The vice president's duties included all planning, design management, budget preparation and overseeing the progress of the various projects being planned. In today's environment the vice president would be labelled the programme manager. Figure 4.5 shows the organisational chart.

4.4.4 Staff
The vice president appointed a development manager who, in turn, formed his team of project managers who were responsible for overseeing each area of work, consisting of the existing hotel renovation and refit, the refit of the utilities plant, the exterior water features, the exterior structures and the landscaping and hardscaping.

The development manager was responsible for coordinating directly with the contractors and hotel operations for all issues related to planning, design, construction, operations, government relationships and general contractor works, as indicated in Figure 4.6.

4.4.5 Milestone dates
Venue A comprises the properties in the Bahamas described above. Venue B is located in the upper east coast of the USA. The plan was to renovate the existing resort then launch the new product

Figure 4.5 Upper management of hotel company

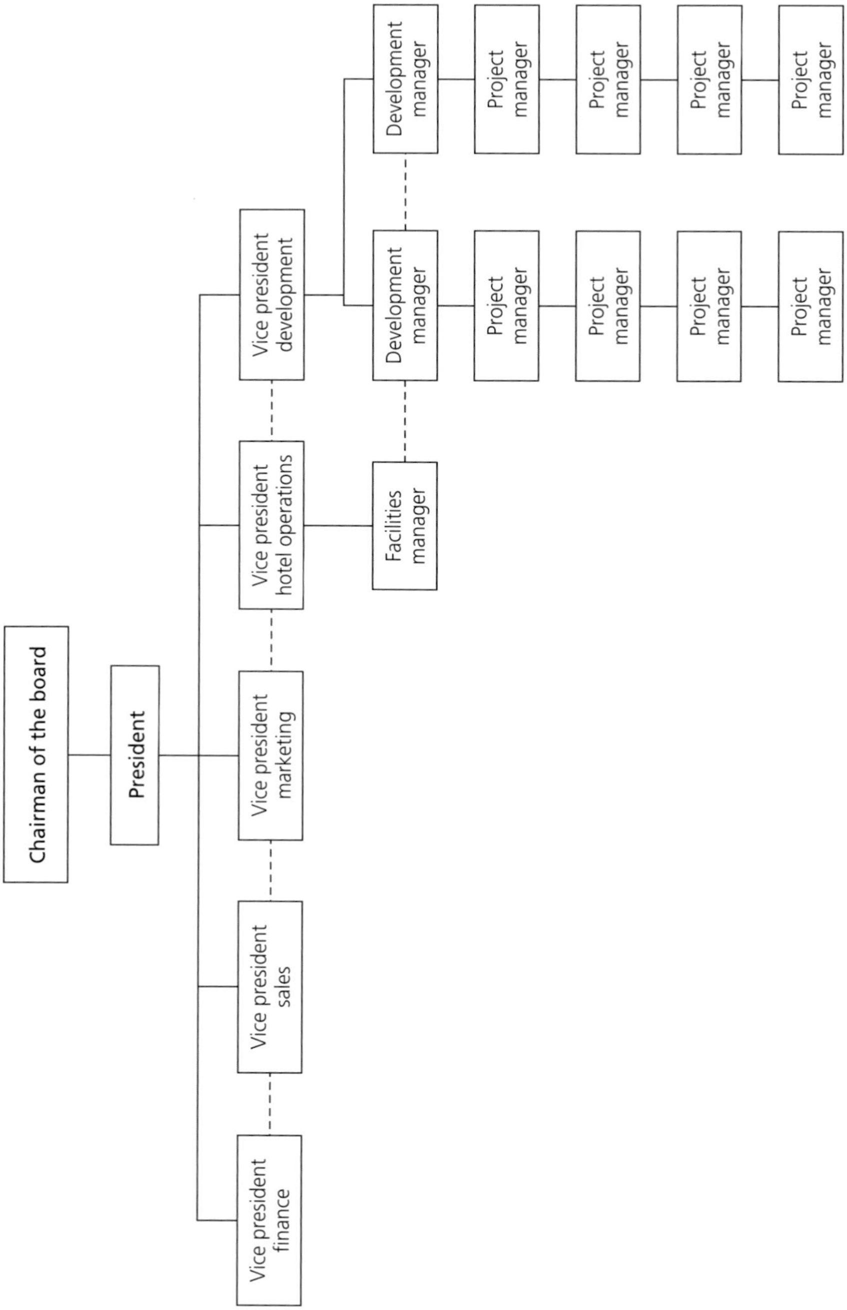

Figure 4.6 Programme management team

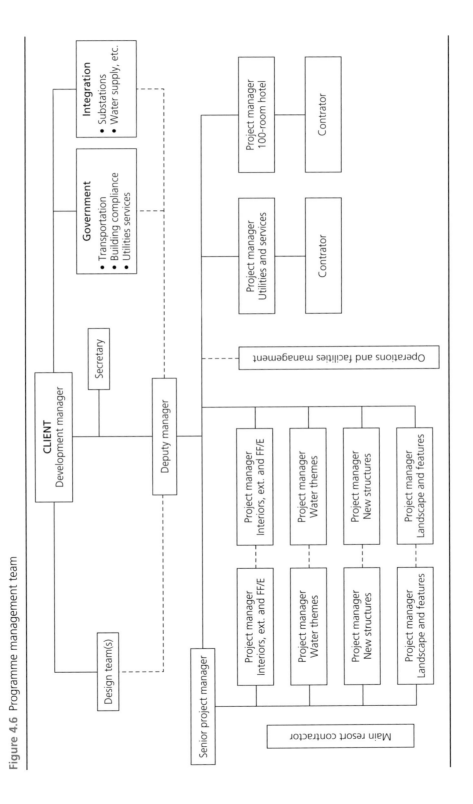

Table 4.5 Milestone dates for hotel company expansion

Activity description	1993	January 1994	April 1994	July 1994	December 1994	1995
Due diligence studies, Venue A	×					
Purchase properties, Venue A		×				
Development company started	×					
Consultant's starts, Venue A		×				
Construction commencements, phase 1			×			
Construction completions, phase 1					×	
Construction commencements, phases 2 and 3				×		
Construction completions, phases 2 and 3					×	
FF & E installation					×	
Reopening under new product name					×	
Due diligence studies, Venue B						×

version in time for the hotel's high season. Therefore the timelines are approximately eight months from commencement to opening. This required daily monitoring of the construction schedule to ensure on-time delivery. Certainly no float existed in this massive undertaking (see Table 4.5).

4.4.6 Pricing and cost control
Due to the fast-track nature of this programme, the most cost-effective contract was a 'cost plus fee' contract with a clause of 'guaranteed maximum price' (GMP). In doing this, the client was assured that the contractors would not exceed a certain price unless change orders were approved. The scope was general in nature and the contract had clauses that locked the contractor to a certain level of intent in respect of the quality, level of materials, systems and finishes. Along with this, there was allowance for items that required additional defining from the design consultants.

4.4.7 Actual cost
The purchase price for the 242 ha with all facilities and amenities was around US$250 million. The construction cost along with the refit of all furniture, fixtures and equipment plus operational supplies and equipment added another US$150 million. Cost of the consultants was approximately US$20 million. Change orders from the general contractor only amounted to US$5 million or only about 7% of the estimate of construction. The total amount of the first phase of the programme was approximately US$425 million. Estimated asset values of the renovated hotel alone exceeded US$600 million after completion. This allowed an expansion to be planned the following year after initial opening and set the company in motion to start planning additional properties.

4.4.8 What went wrong and right
The benefit of a clear vision from the chairman and board of directors allowed the management team to set up a solid basis for the company's global plan of developing the product in many venues. This was accomplished with transparent management of all five of the central responsibilities necessary for a successful programme, listed below:

- plan (scope–schedule–budget)
- risk (financial–schedule–technical)
- procurement
- communication
- conformity.

4.4.9 The benefits of programme management application

The company completed the difficult task of launching a new product in a short time thanks to the following factors:

- The top key managers were involved with all projects, allowing the learned lessons to diminish the risk associated with a global expansion.
- Standardise themed design elements which created efficiencies in design and procurement strategies.
- Programme management had direct communications from top-level management down to the project level.

4.5. Case study: Programme Four
4.5.1 Narrative

A developing country has recognised the need for a large programme to provide housing for its citizen population. These houses will be produced throughout the various regions of the country on a continuous basis up to a finite amount or until the need has been satisfied, and the social benefit realised.

The government of this Middle-Eastern country has decreed the development and construction of 500 000 housing units for its subjects as a first programme. It is planned within an urban community layout, which includes green spaces, asphalt streets, street lighting, underground power, potable water, underground communications networks, hardscaping and landscaping. The task is to plan, develop and construct these urban communities throughout the country in multiple cities within five regions. Due to the magnitude of the overall programme each of the five regions was broken down into separate programmes and assigned to four of the largest consulting companies operating within the country. Section 4.5.2 indicates the size for each programme.

The homes are individual free-standing structures of identical size, shape and layout. The amount of units per community is dependent upon the dedicated land size and topographical features within. Various layouts are designed to create a logical site plan which works for the total area of the land provided.

4.5.2 Data

Total outcome:	500 000 free-standing homes
	450 000 000 m^2 of land
	500 m^2 per home
Programme One output:	45 000 000 m^2 of land
Programme Two output:	20 000 000 m^2 of land
Programme Three output:	30 000 000 m^2 of land
Programme Four output:	18 000 000 m^2 of land

4.5.3   Structure

The magnitude of this programme has demanded the need to engage the services of several programme management companies with similar roles and responsibilities. Each company divided their efforts into two arenas: design and tender and construction operations, as shown in Figure 4.7. The client simultaneously released the programme to four different programme management companies, representing approximately 25% of the 500 000-unit total of homes to be built within the programme.

4.5.4   Staff

Organisation from the client's side consisted of the senior programme director, responsible for communicating progress to the highest ranks of the government and conveying direction down to the programme managers. Reporting to him were two programme managers, each responsible for managing different aspects of the programme; design and tender and supervision of construction.

As counterparts to the client's two programme teams are respective programme teams from the consultant's side for both design and construction supervision for each separate sub-programme, as shown in Figure 4.8.

From the consultant's side, the programme director appointed two regional managers to oversee all construction supervision activities in each respective area. Each regional manager had project managers underneath him on each project. To facilitate good communications each region had a coordinator assigned to link the core team of consultants to the regional teams of the client (see Figure 4.9).

4.5.5   Milestone dates for housing programme

The milestone dates reflected below represent only part of the total amount for Programme One. The design progress was tracked by the four design stages; concept, development, detailed and final. Upon completion of the final design drawings a 60-day period has been allowed for bid tendering followed by a 24-month construction period (see Table 4.6).

4.5.6   Pricing and cost control

Pricing is established during the tender process through preparation of a bill of quantities tied to the tender drawings for each project. The contractor prices each project per the regional efficiencies and difficulties that exist for each project. Upon completion of construction agreements the project costs are monitored by measuring work in place against the established unit pricing defined in the contract and bill of quantities.

4.5.7   Actual cost

Actual cost of the work will be monitored on a monthly basis from work-in-place measurements reported by the consultant. Approximate cost for each home in Programme One was estimated at approximately US$130 000, putting a price tag on Programme One of US$6.6 billion. Added on top of the building cost is that of the infrastructure: another US$4.16 billion. The total cost for Programme One is approximately US$10.8 billion, including soft costs for consultants. The first projects within the programme were in the tendering period at the time of writing. Therefore, details were not yet available.

Figure 4.7 Programme structure

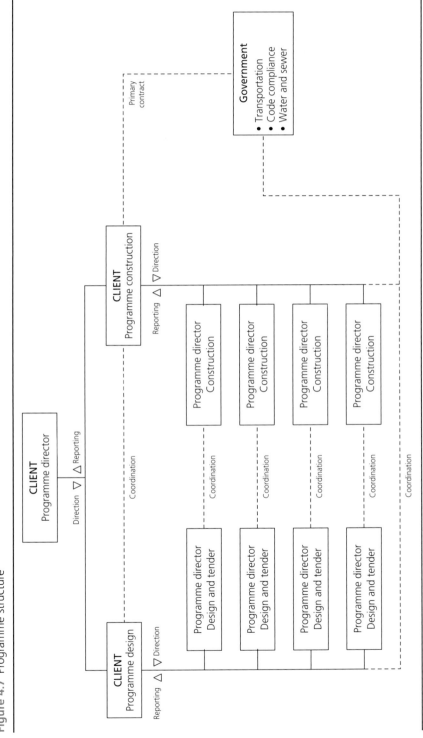

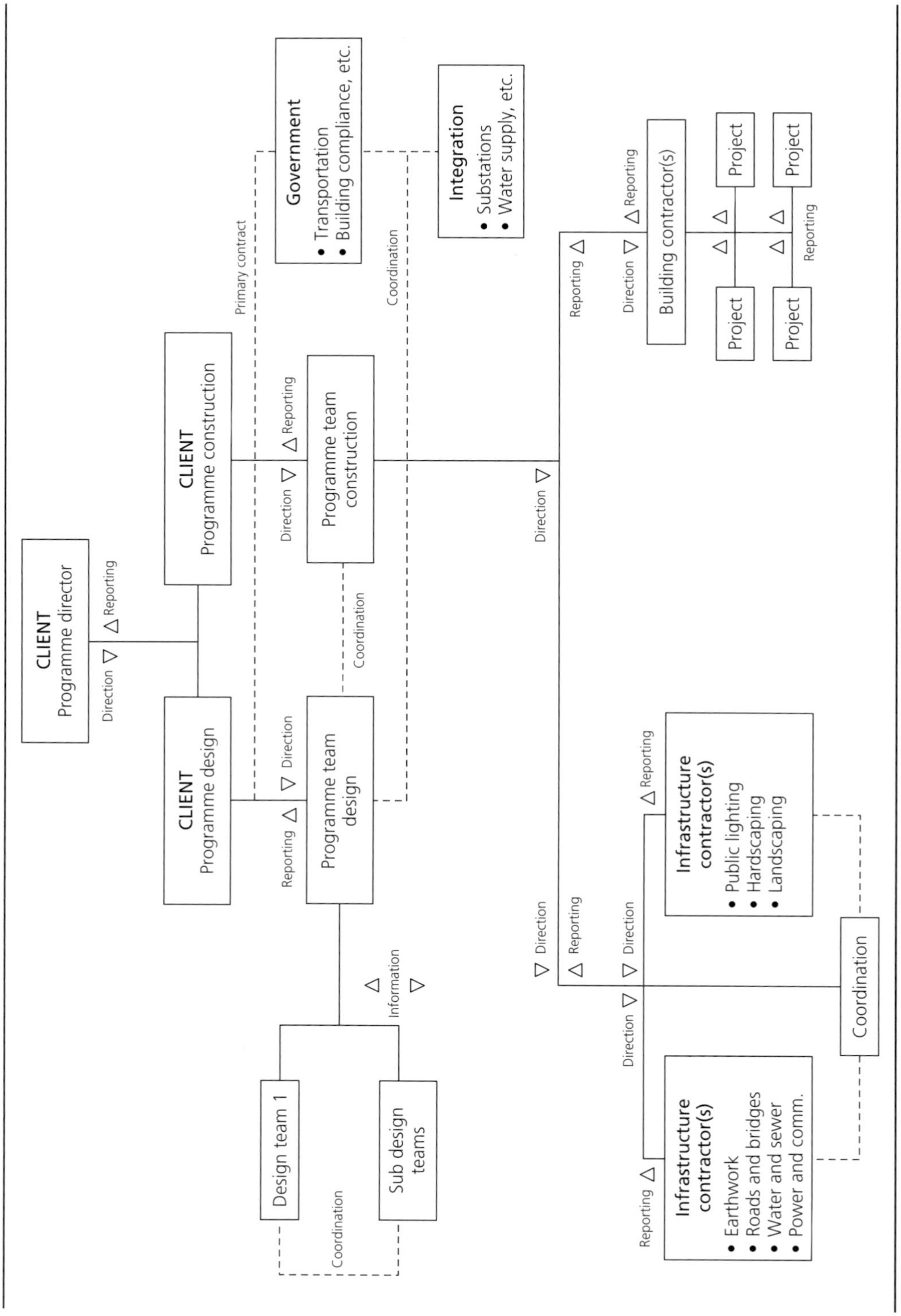

Figure 4.8 Programme management team structure

Figure 4.9 Management team structure

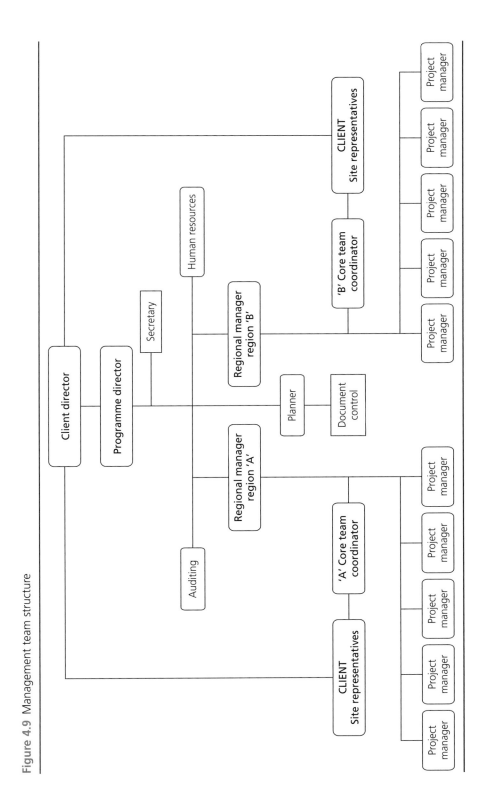

Table **4.6** Milestone schedule for housing programme

			2013							2014											
			Jun.	Jul.	Aug.	Sep.	Oct.	Nov.	Dec.	Jan.	Feb.	Mar.	Apr.	May	Jun.	Jul.	Aug.	Sep.	Oct.	Nov.	Dec.
Package 1																					
Sites	Area m²																				
A	100 000	Stage 1	15 Jun.																		
B	500 000	Stage 2				29 Sep.															
C	600 000	Stage 3							31 Dec.												
D	585 000	Stage 4											15 Apr. Tender	Tender	comm						
E	661 987																				
Package 2																					
Sites	Area m²																				
F	599 984	Stage 1					21 Oct.														
G	400 000	Stage 2							22 Dec.												
H	260 000	Stage 3										23 Mar.									
		Stage 4												15 May Tender	Tender	comm					
Package 3																					
Sites	Area m²																				
I	759 851	Stage 1					21 Oct.														
		Stage 2							22 Dec.												
		Stage 3													16 Jun.						
		Stage 4																24 Sep. Tender	Tender comm		

4.5.8   What went wrong and right

At the time of writing, the structure of management from the client's side is missing several key elements that will create poor results due to management not having access to sufficiently high-quality information to allow them to make good decisions. These elements are as follows:

- issue: lack of programme execution plan
  - risk: not meeting the plan for schedule, financial and technical issues
- issue: task management system
  - risk: poor communication
- issue: missing global documents control system
  - risk: communication and conformity
- issue: central procurement strategy
  - risk: non-conformity and volume-pricing discounts

4.5.9   The benefits of programme management application

The magnitude of this project demands a proper programme management model to ensure success through sound techniques, processes and procedure. Without a transparent management model where all stakeholders have an agenda that is focused on the project and not on the personnel, the chances of success are poor and the chances of risk are high.

4.6.   Summary
4.6.1   What went wrong and right

Every organisation or entity has a different approach to the set-up and operation of a programme and/or a project. The programme manager or project manager can be defined in many ways, and can have many roles outside the programme.

The fatal flaw in any programme is a lack of planning early in the lifecycle and an under-estimation of the resulting issues that may occur without complete stakeholder buy-in at the very beginning of the cycle.

Without a doubt the most important event at the beginning of the programme is to define the key client goals, understand the programme needs and set milestone dates to measure progress. From this point the key decision makers need to be appointed early on. Their roles, responsibilities and limitations should be defined and clearly understood by all programme participants. As well, it is essential to finalise the scope and designs before commencement of any project to avoid future changes, which create delay and cost overruns.

All the case studies are examples of the client and stakeholders having a clear vision of their main goals; development and construction of buildings, creating long-term outcomes for their organisations from the sales of real estate, expansion of portfolio assets or the benefits of social advancement. Programme One managed the changes in the market and maintained a positive outcome for the company on a long-term basis. Programme Two has faltered due to the lack of planning and implementing the fundamentals of management for programmes and projects. Programme Three was successful do to their existing solid management structure that maintained involvement in all projects applying the benefits of lessons learned, common design components and common work packages. Programme Four identified the troubles that will ensue when a lack of a clear organisation and decision making is missing from the client, creating an absence of good communication flow throughout the programme team.

4.6.2 The benefits of programme management application

The benefits of applying a programme management approach can be summarised as follows:

- informed decision making
- clear direction downward
- predictable delivery of goals
- fast problem identification and solving
- transparent communication
- elimination of personal agendas
- cross-programme benefits and efficiencies.

REFERENCE

Flyvbjerg B (2006) Five misunderstandings about case study research. *Qualitative Inquiry* **12(2)**: 219–245, **doi**: 10.1016/j.ijproman.2012.01.020.

Programme Management in Construction
ISBN 978-0-7277-6014-2

ICE Publishing: All rights reserved
http://dx.doi.org/10.1680/pmic.60142.079

Chapter 5
Programme management contracts formation

5.1. Introduction

This chapter will look at the various types of contracts that are, or may be, used in the programme management process, and will consider the suitability of those various contracts in different circumstances. It will also consider how those contracts may need to be amended and/or developed to suit the requirements of the programme management process and the need, in applicable circumstances, for a more integrated procurement process to be used. There are obviously many factors that affect the choice of contracts to be used when procuring a construction programme; but the main points, as always, generally come down to the question of quality, price, time and the size and number of projects within a programme.

It is the balance between the sometimes competing issues of quality, time and price that normally establishes the requirements of the client, and this is usually a key driver for the choice of the procurement approach and of the contract to be used.

In the 20th century, procurement was considered largely for the construction element of a project, but now in the 21st century, and in the programme management process specifically, a more holistic approach is normally taken to procurement. In this approach, the capital and/or construction phases of a project (or projects) are not considered in isolation from the lifecycle of the project (or projects). This approach is often referred to as 'sustainable procurement', which means that the procurement of works, services and goods relies on the making of decisions that relate to the entire lifecycle of a programme.

Obviously, before an appropriate procurement approach can be decided on, it is necessary for the programme's objectives to be established. These objectives can only really be established from a consideration of the client's required optimum combination of whole lifecycle quality, price and time.

This may seem to be obvious, as surely all clients want a facility to be top quality, to be at a low cost, and to be in use as quickly as possible.

However, this is not always the case: the client's budget will not always allow for its ideal requirements in respect of quality and time, and compromises may need to be made in respect of these matters. Alternatively, it may be that one or more aspects of the quality–cost–time triangle are not that important to a client in a particular situation. Therefore, for example, it is not always the case that a client requires a programme's facilities to have longevity, and in some instances the programme's facilities may only be constructed for a particular purpose or event and after that purpose or event has passed the programme facilities may need to be removed. Therefore, in such a case, the long-term quality of the programme facilities may not be as important as the speed and cost in providing the programme facilities in the first place. Thus, this objective of speed and cost, being more important than quality, would have a major impact on the choice of the procurement strategy to be used.

Therefore, generally, the procurement strategy should identify the best way of achieving the programme objectives and this strategy will determine, among other things, the level of integration of the design, construction, maintenance and operation for a series of projects within a programme.

The procurement route that is chosen is the method that is decided on to achieve the required procurement strategy.

Obviously, in advance of the procurement route being decided on, there needs to be a contract between a client and the programme management consultancy practice in addition to the contractor(s) executing the construction of the multiple projects within the programme.

5.2. The contract between a client and a programme management consultancy practice

The contract between a client and a programme management consultancy practice is normally an ad-hoc form which sets out:

- the names of the parties (that is, normally the client and the programme management consultancy practice)
- the services to be provided by the programme management consultancy practice – services which may include for the planning, budgeting, scheduling, expediting, coordinating and managing of the process to ensure that the project objective is met
- the limitation of liabilities of the programme management consultancy practice when acting as an agent of the client
- details of any indemnity provided by the client to the programme management consultancy practice in respect of debts, claims and liabilities incurred by the programme management consultancy practice in the performance of its functions under the contract
- the fee (or the fee structure) payable to the programme management consultancy practice for providing the required services; and the manner in which that fee will be paid.

This contract could be loosely based on the Fédération Internationale des Ingénieurs-Conseils (FIDIC) 'Client/Contractor Model Services Agreement' (the White Book); or it could be based on any other of the multitude of standard consultancy agreements or ad-hoc forms that serve this purpose.

5.3. Procurement routes

Depending on the procurement strategy decided on, there are many procurement routes available for use in the programme management process. These include:

- the traditional approach
- design and build
- design, build and operate
- design, build, operate and transfer

- design, build, operate and maintain
- single purpose entities
- joint ventures
- consortia
- partnering
- collaboration
- project alliancing.

These procurement routes may be applied by way of a standard form or on an ad-hoc basis; and these routes are considered further below.

5.3.1 The traditional approach

The traditional approach in the construction industry is to have design as a completely separate function from construction.

In this traditional approach a client would have one contract with the designer (for the design) and an entirely separate contract with the contractor(s) (for the construction of the programme), but there would be no direct contractual link between the designer and the contractor(s) (see Figure 5.1).

Figure 5.1 The traditional procurement route

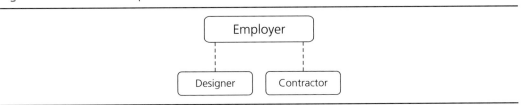

The contract between the client and the contractor could be, for example, the FIDIC *Conditions of Contract for Construction* contract (the Red Book) (FIDIC, 1999b); or the *JCT Standard Building Contract* (JCT 2011a), or any other similar standard form contract that does not involve the contractor in design.

When using the traditional approach, there is usually more certainty of price (particularly as the work should be fully designed in advance). However, with regard to quality, while the client retains responsibility for and has control of the design team and there is direct reporting by the design team to the client to ensure that quality control is maintained, there is the potential for over-design and/or over-engineering and the contractor has little input in the design process. In respect of time, to be effective, the 'traditional' approach requires the scheme to be more or less fully designed before tenders are sought and this often results in an extended pre-tender period. If the design is not completed before tenders are sought this can lead to the 'design as being built' approach being adopted, which invariably delays the on-site programme period.

5.3.2 Design and build

In a design and build contract, the contractor acts as the single point of responsibility to the client for the design, management and delivery of a project, in accordance with the required

Figure 5.2 The design and build procurement route

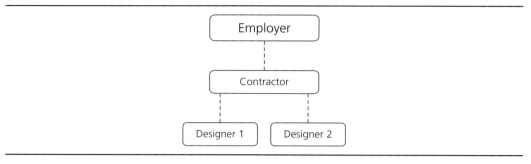

specification, the required time period, and the required price of each project within the programme (see Figure 5.2).

Design and build contracts (and other similar arrangements) solve the boundary problem between the various design consultants by in effect increasing the boundary's perimeter until it absorbs all of the key participants. Thus, information sharing and reliance issues are resolved by joining the provider to the other party.

Design and build is a project delivery method that combines two, usually separate, services in a single contract. With design and build procurement, clients execute a single, normally fixed-fee contract for both the design services and the construction for the whole programme. This, however, in extreme circumstances, can be divided into two, three or a few contracts to break the programme into manageable sub-programmes each consisting of a limited number of projects reflecting the capacity of the designer and/or the contractor.

A typical example may be the JCT *Design and Build* form (JCT, 2011b).

For this strategy to be fully effective, the key participants must be identified and included in the design and build team. This is automatically accomplished if the designers are directly employed by the design and build contractor, but it is more challenging if the designers or key systems providers are subcontractors to the design and build contractor.

With design and build contracts, the design and build contractor assumes responsibility for the majority of the design work and all construction activities, together with the risks associated with providing these services for an agreed (normally fixed) fee. When using design and build delivery, clients usually retain responsibility for financing, operating and maintaining the programme.

However, even with design and build procurement, the client must usually complete a certain amount of preliminary engineering and programme definition in order to be able to prepare tender documents for each project within the programme or for the whole programme. A programme that is too advanced (for example, fully designed) may be unattractive since there will be minimal opportunity for the private sector to apply innovative methods to reduce cost and schedule. On the other hand, a programme that still is at an early stage with unanswered questions regarding scale, alignment and other project features will be difficult to structure on a design and build basis because the potential private sector contractor will be unable to reliably assess schedule and costs, and may build in an excessive and unpalatable risk factor.

5.3.3 Design, build and operate (DBO)

In a design, build and operate (DBO) contract, the contractor carries out the duties of a design and build contractor as set out above, but in addition operates the facility over an agreed compliance period, primarily to prove the contracted assumptions.

In such a case, the contract between the client and the contractor could be, for example, the FIDIC 'Design, Build and Operate Projects' contract (the Gold Book) (FIDIC, 1999a).

5.3.4 Design, build, operate and transfer (DBOT)

A design, build, operate and transfer (DBOT) programme is typically used to develop a discrete asset rather than a whole network and is generally entirely new or greenfield in nature (although refurbishment may in some cases be involved).

In a DBOT programme, the managing company or operator generally obtains its revenues through a fee charged to the client rather than from tariffs charged to consumers. Some projects are called concessions, such as toll road projects, which are new build and have a number of similarities to DBOTs. This is a type of arrangement in which the private sector builds an infrastructure programme, operates it, and eventually transfers ownership of the programme to the client. In many instances, the client (which may be a utility or a government) becomes the contractor's only customer and promises to purchase at least a predetermined amount of the programme output. This ensures that the DBOT contractor recoups its initial investment in a reasonable timespan.

5.3.5 Design, build, operate and maintain (DBOM)

In a design, build, operate and maintain (DBOM) programme, the contractor carries out the duties of a design, build and operate contractor (as set out above) but in addition maintains the programme facilities over an agreed period of time (normally over 10 years, although this can be a much longer period if agreed, and depending on the size and complexity of the programme).

The DBOM approach is an integrated partnership that combines the design and construction responsibilities of design–build procurements with operations and maintenance. These programme components are procured from the private sector in a single contract with financing secured by the public sector. With a DBOM contract, a private entity is responsible for design and construction as well as for long-term operation and/or maintenance services. The public sector secures the programme financing and retains the operating revenue risk and any surplus operating revenue. The advantage of the DBOM approach is that it combines responsibility for usually disparate functions – design, construction, and maintenance – under a single entity. This allows the contractor to take advantage of a number of efficiencies. The programme design can be tailored to the construction equipment and materials that will be used. In addition, the DBOM team is also required to establish a long-term maintenance programme up front, together with estimates of the associated costs. The team's detailed knowledge of the programme design and the materials utilised allows it to develop a tailored maintenance plan that anticipates and addresses needs as they occur, thereby reducing the risk that issues will go unnoticed or unattended and deteriorate into much more costly problems.

5.3.6 Single purpose entities (SPE)

Single purpose entities (SPE) also solve the boundary problem by bringing all parties within one boundary. An SPE is a limited liability enterprise (corporation, limited liability company, limited liability partnership, for example) created to design, construct and possibly own and operate a programme facilities.

The key participants sponsor the SPE and achieve a gain by optimising the SPE's success. The SPE contracts with the client for the services required to construct the facility, with the detailed requirements of scope, responsibility and liability determined on a programme-specific basis.

The parties within the boundary of the SPE must release each other from most potential liabilities or agree that any 'in boundary' claims will be paid for only by way of specific project insurance.

SPEs are common in off-balance-sheet-asset-financed programmes (programme finance). Under a classic programme finance structure, non-recourse loans are used to design and construct a revenue-generating asset that is owned by the single purpose entity. The asset, and any guaranteed income streams, secures the loans. The fundamental economic principle of the SPE is that the client's return is based on creating value in the SPE.

The SPE is not often a suitable contract for the execution of large programmes comprising many entities. It can be used, however, when the programme is confined to a certain and defined geographical location and can be delivered using local resources and a contractor familiar with the programme to be constructed.

5.3.7   Joint ventures (JV)

A joint venture (JV) is a business agreement in which the parties agree to develop, for a finite time, a new entity and new assets by contributing equity. The parties exercise control over the enterprise (the JV) and consequently share revenues, expenses and assets.

The JV can be created for one specific programme only – in which case the JV is referred to more correctly as a consortium (which was the case in the building of the Channel Tunnel) – or may be for a continuing business relationship. In either case, the JV is normally dissolved when the goal that it was set up for has been reached.

A JV takes place when two parties come together to take on one programme. In a JV both parties are equally invested in the programme in terms of money, time, and effort to build on the original concept. JVs are used by large companies and organisations to diversify or carry out programmes that may normally be outside of their scope. A JV allows both parties to share the burden of the cost of the programme, as well as the resulting profits. A good example can be the construction of a programme constituting a number of electrical substations where a contractor specialising in building works teams up with a contractor specialising in electrical works.

Since financial cost and return is involved in a JV, it is necessary for a strategic plan to be in place. Both parties must be committed to focusing on the future of the partnership, rather than just the immediate returns; while recognising that both short-term and long-term successes are equally important. In order to achieve this success, honesty, trust, integrity and communication within the JV are obviously essential.

5.3.8   Consortia

Construction consortia arise in several different forms. They are one of many ways in which traditionally separate parts of the construction procurement process might be integrated.

Sometimes, this integration involves a single firm taking on obligations wider than it is capable of undertaking alone, and then subcontracting elements of the programme; at other times, groups of firms get together to act as a consortium to meet the needs of a client.

Construction firms are often encouraged to work in teams and at the same time involve private sector finance in the funding process through some form of Private Finance Initiative (PFI) approach. In response to this approach, Special Purpose Vehicles (SPVs) are normally set up to structure the delivery once a contract has been agreed.

Sometimes, before a contract can be agreed, informal consortia may be formed, combining banking, property and construction companies. Only when one of these informal consortia wins a bid is an SPV formally established in respect of the project in question.

A good example of a consortium is the construction of railway networks. Each company in the JV or consortium can be familiar with a scope such as the stations, the railtrack, the signals and the trains.

5.3.9 Partnering

A partnering agreement involves a number of firms, normally including the client, working co-operatively to achieve a given output over one or a number of projects within a programme (see Figure 5.3). Remuneration is usually based on contract terms and sometimes in addition on some form of incentive payment which may be based on the contribution by a partnering party to the success of the programme (normally in terms of time or financial return).

Partnering can be either long-term for a specified period (often referred to as strategic partnering) or programme-specific; however anecdotal evidence trends show that strategic or long-term partnering more often provides greater opportunity for improvement and benefit.

Partnering is intended to be a flexible approach which is not constrained by the rigid contractual barriers between a client and the suppliers of services. Partnering connotes the way in which parties agree to conduct themselves, and the behavioural attitudes that will be adopted by them.

Partnering can involve any or all of the designer consultants, client, main contractor and/or contractors, and it is an arrangement based on openness, integration and collaboration.

Figure 5.3 Partnering agreement

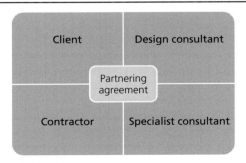

The theory is that when partnering takes place the relationship between the parties changes from being linear, up and down the contractual chain, to being based on a desire to solve issues on a more equal footing. This might mean that the whole supply chain becomes involved in management decisions, and there tend to be high-level joint objectives between the partnering partners and also shared rewards for reaching those objectives.

More often than not, partnering does not replace the need for a contract to define the legal basis of the relationship between the parties, but simply supplements those contracts with an overriding partnering ethos.

Partnering may be achieved through a number of different approaches, although it is important that the parties decide if any agreed partnering principles are to be legally binding or not. If they are to be non-binding, then the language used throughout the partnering documentation must clearly reflect that intention to avoid any doubt.

In terms of overall approach, the parties might opt for either 'strategic' or 'programme' partnering.

Where the parties choose 'strategic' partnering, a long-term relationship is developed between any or all of a main contractor, client, designer consultant or subcontractors to the benefit of each of the parties over the course of a series of projects. These parties will then work together to meet agreed targets which will have been formalised in a binding or non-binding partnering agreement. There will not necessarily be a guarantee of specific work.

'Programme partnering', in contrast, comprises free-standing binding or non-binding 'partnering charters' for single projects within a programme.

Beyond the two overarching approaches listed above, there are a range of contractual approaches to partnering:

- Framework agreement: an umbrella agreement which sits on top of, but does not interfere with, the underlying contracts.
- Multi-party binding contract: each party to a multi-party contract is in a direct contractual relationship with every other party.
- Construction consortium: parties might together form a consortium to tender for a specific project or projects, often relating to major infrastructure projects. This might be, for instance, where the different parties have different specialist skills all of which are required for a successful tender, and by working together the group will either improve its position in the market or win a series of projects.
- Alliance: usually a formal partnering structure and generally used only on major projects. Alliancing in its simplest sense is bringing together two or more businesses and joining them together to achieve a common goal. This may be a formal joining (e.g. setting up a specific company) or something else. There are no hard and fast rules.
- A non-binding 'partnering principles' document.
- Bolt-on clauses. These are not otherwise attached to a partnering arrangement but can be added to a standard form contract as an additional extra. The idea is that all parties add the bespoke clause so that everyone has an element of partnering to their otherwise 'normal' contracts.

Partnering is often favoured by clients wanting to build complex and specialist programmes including technology as a key part of the programme – this includes telecommunication, airports, desalination plants and water treatment plants.

5.3.10 Collaboration
Collaborative working is a process undertaken by two or more organisations sharing their collective skills, expertise, understanding and knowledge to jointly deliver the best solution that meets the programme objectives.

Collaborative working on projects is increasing within the construction industry and is used frequently in the programme management process.

5.3.11 Programme alliancing
The collaboration approach often leads onto project and programme alliancing.

Programme alliancing may be defined as a commercial and/or legal framework between a client and one or more parties for the delivery of a project.

Programme alliancing is characterised by the following features:

- collective sharing of the programme risks
- no fault, no blame and no dispute between the alliance participants (except in very limited cases of default)
- foresight applied collaboratively, which mitigates problems and shrinks risk
- clear division of function and responsibility, which helps accountability and motivates people to play their part
- payment of parties providing services under a three-limb payment plan comprising
  - the reimbursement of the party's project costs on a 100% open book basis and/or on a more front-loaded basis
  - a fee to cover corporate overheads and normal profit – a gain/pain share process where the rewards of outstanding performance and the pain of poor performance, and the financial gains and losses, are shared equitably among all alliance participants
  - unanimous principle-based decision-making on all key programme issues
  - an integrated project team selected on the basis of the best person for each position.

Two common alliancing features work particularly well with BIM:

- First, in an alliancing project, the parties agree that they will not sue each other, except for wilful default. Sharing information cannot lead to liability, therefore the liability concerns that may impede BIM adoption do not apply in an alliancing project.
- Second, because a portion of compensation is tied to a successful outcome, there is an incentive to collaborate. In this context, BIM is an ideal platform for interactively sharing information, ideas and solutions.

A secondary but important theme of alliancing is that people will be motivated to play their part in collaborative management if it is in their commercial and professional interest to do so.

Collaborative project approaches are beginning to appear in the USA under the general term integrated project delivery (IPD). Unlike alliancing, which has generally been used for major civil and industrial infrastructure, IPD in the United States has been primarily used for complex structures, such as hospitals.

IPD is a radical departure from traditional prescriptive and adversarial contract approaches and it offers the potential for increased programme value and greater reward for all participants to manage uncertainty and risk, thus eliminating the fear that causes participants to focus on their narrow self-interests.

5.4. A building information modelling (BIM) approach to contracts

Given the rapid advances in technology (and in particular the increased use of building information modelling or management (BIM)), it appears to be inevitable that collaboration will be required to be applied more and more over the coming years.

At its most basic level, BIM is a computerised process which is used to design, understand and demonstrate the key physical and functional characteristics of a building (or a construction or civil engineering project) on a 'virtual' computerised model basis. BIM, therefore, provides the opportunity to concurrently design and visualise the building in 3D, although it needs to be understood that the opportunity to visualise in 3D is a product of BIM, and not the process of BIM itself.

At its more advanced levels, being something that is bound to happen more and more in the future, BIM is the use of a computer software model to simulate the construction and operation of a building. The resulting model, a building information model, being a digital representation of the building from which views and data appropriate to various users' needs can be extracted and analysed to generate information that can be used to make decisions and improve the process of both delivering the building and for the entire lifecycle use of the building.

When applied correctly, BIM is intended to make substantial cost savings throughout the whole lifecycle of a building, from design, through construction, through its maintenance, to its regeneration and eventual disposal or recycling.

In respect of the design and construction phase, the BIM process, when fully applied, possesses the potential to save valuable resources, including time, money and natural materials. BIM will do this by reducing the amount of inaccurate and conflicting information, which, in consequence, will reduce variations, alterations and delays; and this will be achieved by the building being constructed in the 'virtual' world, before it is built physically.

The adoption of the BIM process during the design and construction phase of a building by those parties interested in the successful completion and outcome of the building, from initial design to practical completion of the property and beyond, offers the opportunity to achieve systematic co-ordination and use of all data available.

At the time of writing, the construction and civil engineering industries are on the threshold of seeing digital information flows from inception through to demolition of projects within the construction and civil engineering industries, which is bound to result in large efficiencies being achieved in many areas.

Another major aspect of BIM is the potential full collaboration of the entire project team – the client, the architect, the engineers, the consultants, the contractor and the specialist contractors – in developing the project design. This full collaboration not only allows for the increased speed of project delivery, enhanced economics for the project, and true lean construction at all levels; it also has the potential to change the relationships between the participants in the construction industry from traditional contracts based on obligations and rights to the more modern partnering associations based on a fair allocation and sharing of risks and liabilities.

BIM technology can of course be used solely to produce better-quality design documents without any intent to share information or to use the more extensive functionality that BIM allows. When used in that limited way, BIM is simply an advanced form of computer-aided design (CAD) that adds very little to the existing process.

However, in order to optimise efficiencies from a process such as BIM, a collaborative team structure needs to be in place. That team structure needs to be one in which team members are either contractually obliged to, or in some other way have agreed to, work in a unified manner; and the structure needs to be one where the team members will provide each other with data that will allow the other team members to perform their work faster, better and/or cheaper. At the same time, each team member needs to be able to insert, extract, update or modify information into the BIM model to support and reflect its own role in the process, and to ensure that each project member remains responsible for its own element of the design and/or the data input (or not, if that is what has been agreed).

Used in this way, BIM serves as a catalyst to change for the relationships between the parties and eventually for the basis of their agreements.

Collaboration through BIM is a profound change that creates great opportunities as noted above, but also creates new legal and liability issues.

All significant construction projects incorporate a written contract at their core. Contracts legally record that which has been agreed on by the parties to the contract, but not all agreements are legally binding: there are essential elements that are necessary to exist, and some that must not exist, for a contract to be constituted. Beyond that basic description, construction contracts are as complex as those matters that are both within, but also beyond the original anticipation, yet within the original contemplation, of the parties to the contract.

It is what construction contracts oblige stakeholders to do or refrain from doing, either expressly or implicitly, and it is those matters upon which the contracts remain silent completely, that has formed the basis of construction contracts often being accused of being adversarial.

Contracts determine the parties' risks and rewards, but also the rights and obligations not only of the parties, but also sometimes of wider stakeholders too. In doing so, the construction contract creates a legal framework upon which the stakeholders' duties and obligations are defined.

Over many years past, a major concern has been that the adversarial nature of the construction industry has been cultivated, in part, by contracts that are viewed as being confrontational in nature and which can quickly escalate disagreements to disputes.

Traditionally, most contracts have been bipartite agreements (that is, an agreement between two parties only) with clearly defined boundaries in respect of scope and liability. These contracts tend to insulate and isolate rather than collaborate. One contract tends to define one pair of relationships or one set of requirements in a programme, each having little regard for the other.

Many of the most fiercely fought battles in construction law focus on the dividing line between entities. Privity of Contract, the economic loss doctrine, means and methods, and third-party reliance are all issues where drawing lines between parties is essential to determining responsibility and liability.

The consensus of opinion appears to be that the use of BIM at Level 2 (i.e. a managed 3D computer environment) does not require wholesale changes to the traditional forms of contract or the allocation of responsibilities as between the parties. However, in contrast to this, Level 3 BIM (a fully integrated and collaborative computer process) utilises a single project model, accessible by all team members.

Therefore, as BIM moves towards Level 3 in the future, changes to building contracts will almost certainly be necessary, as the traditional legal position and the relationship between the parties are likely to change.

The Level 3 of BIM maturity will raise significant contractual, legal and insurance issues, including:

- the priority of contract documents, e.g. between the contract conditions and the BIM protocol
- multiple model relationships and conflict prioritisation, including data derived from same
- design liability and thereby professional indemnity insurance issues
- matters concerning intellectual property rights.

Many of the issues described above are in connection with collaboration (which will be required if BIM is to operate efficiently and effectively) and relate to duties and obligations that transcend boundaries between normal bipartite contracts.

As the use of BIM further develops to the level of fully integrated BIM, it may be that bipartite contracts, even with BIM addendums or protocols, may become unsuitable, and collaborative multi-party contracts could potentially become more appropriate. The BIM process has the potential to substantially alter the relationships between parties and blend their roles and responsibility. Risks will need to be allocated rationally; based on the benefits a party will be receiving from BIM, the ability of the party to control the risks, and the ability to absorb risks through insurance or some other means.

There will be a need for standard contract documents suitable for BIM, and without these the development of BIM will almost certainly be hindered.

Standard contract documents perform three key functions:

(a) They validate a business model by providing a recommended framework for practice.
(b) They establish a consensus allocation of risks and an integrated relationship between the risks assumed, compensation, dispute resolution and insurance.

(*c*) They reduce the effort involved in documenting the roles and responsibilities on a programme.

Unfortunately, the emergence of BIM as a vehicle for dramatic change in design and construction occurs in a legal environment that has not fully come to grips with all the risk management implications of the underlying technology of electronic representation, or transmission of documents of any type.

It may be said that liability only requires that there be intent to influence and reach a group or class of persons, and therefore contractors and subcontractors relying on the design within a model may be able to bring an action against the designer for damages caused by negligent errors.

However, in a collaborative programme, the designer is aware that other parties are relying on the model's accuracy; and it is a very short step from foreseeability, to knowing that the model is intended to provide information for the contractors' and subcontractors' benefit.

Against this background, it is clear (particularly as the process develops further) that BIM is essentially collaborative; and it will be most effective when the key participants are jointly involved in developing and augmenting a central BIM model.

Therefore, although roles and duties will remain, the transitions and boundaries between the various team members will be less abrupt and less easily defined.

There is clearly a tension between the need to tightly define responsibilities and limit reliance on others, and the need to promote collaboration and encourage reliance on information embedded in a BIM model.

5.5. Main elements of a programme management contract

This part of the chapter sets out the main elements that need to be included in a contract for a programme.

- The law of the contract: the substantive law of the contract must be stated.
- The language of the contract: the language to be used for all communications between the employer, the engineer and the contractor must be stated.
- The contractor's legal entity: the legal entity of the contractor needs to be stated. This may be a legal partnership or company and/or may be individuals or partnerships.
- The location of the works: the geographical location of the various projects making up the programme of works must be set out. If the programme is to be broken down into sub-sections, this needs to be stated.
- The scope of the works: the scope of the works to each programme and/or to each project within a programme should be set out in detail.
- Design liability: the extent of and level of design liability of the contractor needs to be stated.
- The programme or project duration: the durations for the programme as a whole and for the individual projects within a programme need to be stated.
- The schedule of operations: the requirements for and contents of a schedule of operations for an individual project or for the programme as a whole need to be set out in the contract. The schedule of operations needs to be subject to the approval of the employer or engineer

and must consist of detailed timetables which cover the programme entirely, in addition to detailed timetables for each project individually. The schedule of operations needs to outline the proposed procedures to be applied to execute the works and needs to show the critical path through the programme. The contractor should also be required to submit written statements that include a general description of the proposed arrangements and method statements for the execution of the works, and the contractor needs to be required to submit updates of the schedule of works with each application for payment made.

- Liquidated damages: the level of liquidated damages for delay (normally charged on a weekly basis) must be stated.
- The contract price: the contract price (for either the programme as a whole or for the individual projects) needs to be stated, and a contract sum analysis in respect of the contract price needs to be provided by the contractor. It needs to be made clear that the contract price and the contract sum analysis is deemed to include all the costs which the contractor may incur to meet all the obligations defined in the contract. That value is also to include all the other work items as detailed in the technical specifications, drawings, bill of quantities or any other contract document defined in the contract.
- Periodic payment: the procedure and timing of periodic payments for the programme need to be set out. Any approval and/or verification process for payments must also be clearly stated.
- Extension of time: the rights for a contractor to obtain an extension of time and the process that a contractor needs to follow to be granted an extension of time must be set out. Any condition precedent clauses in respect of an extension of time must be stated.
- Monthly reports: the requirement for a contractor to submit a monthly report and to include a work progress schedule for the execution of each programme activity, together with the percentage completed for each of those activities, needs to be included.
- Site record: a requirement for a daily written site record that details the contractor's works needs to be stated.
- Work management and coordination with existing utilities: an obligation on the contractor to schedule and complete the works to meet the least possible interruption in the regular and continuous flow of works of the services networks should be included.
- Coordination and supervision of subcontractor activities: a requirement for the contractor to coordinate and supervise the works of all other subcontractors to an extent that allows the execution of the smooth flow of work without contradiction between disciplines and in such a manner that no discipline at any time will delay the progress in general, needs to be incorporated.
- Value engineering, designs and alternative solutions: a requirement that the contractor shall submit value engineering studies in order to evaluate any of the project elements to meet the optimal suggestions or alternatives which achieve the project purposes at the highest quality levels and the lowest cost, in the shortest time, or with the fewest modifications to any systems or elements of the project, needs to be incorporated.
- Quality control by the contractor: an obligation on the contractor to provide a quality control system that is based on sufficient tests and recording of all the work items including the activities of subcontractors to ensure compliance with specifications and applicable drawings in terms of material and workmanship, structure and functional performance, must be included.
- Technical support and backup: the contract should include a requirement that the contractor is to provide the required engineers, quantity surveyors and admin staff to support the employer's staff.

- Protection of works, properties and individuals: a requirement that the contractor shall procure the methods, materials and structures required to ensure that losses in the works, materials, properties or lives shall not occur due to the contractor's performance of the works under the contract needs to be incorporated.
- Contractor to clear and clean the site: an obligation on the contractor, during the work execution, to evacuate from the site all obstacles that should not be present, and to store or dispose of any construction equipment or any surplus material and clear the site and dispose the debris and temporary works which are not needed, needs to be included in the contract.
- Work insurance: the insurance requirements for the programme and/or for the individual projects needs to be set out in the contract.
- Typical appendices to be attached to (and incorporated within) the contract are as follows:
  - Appendix A drawings: architectural sketches, structural sketches, mechanical drawings and electrical drawings
  - Appendix B specifications
  - Appendix C bill of quantities
  - Appendix D instructions for architectural works, structural works, mechanical works and electrical works
  - Appendix E location of works
  - Appendix F soil investigation report.

REFERENCES

FIDIC (1999a) *FIDIC Conditions of Contract for Design, Build and Operate Projects* (*DBO Contract*) – *The Gold Book*. FIDIC, Geneva, Switzerland.

FIDIC (1999b) *FIDIC Conditions of Contract for Construction – The Red Book*. Thomas Telford, London.

JCT (2011a) *JCT Standard Building Contract With Quantities 2011: (SBC/Q)*. Sweet & Maxwell, Andover, Hampshire.

JCT (2011b) *JCT Design & Build Contract 2011 (DB)*. Sweet & Maxwell, Andover, Hampshire.

Programme Management in Construction
ISBN 978-0-7277-6014-2

ICE Publishing: All rights reserved
http://dx.doi.org/10.1680/pmic.60142.095

Chapter 6
Planning in programme management

6.1. Network-based scheduling – an introduction

Network-based scheduling techniques and programming of the complex sequences of activities, and their dependencies for a typical programme, offer an invaluable tool not only for planning and scheduling but also for negotiating timely settlement of variations, disputes and delays throughout the life of the construction programme. It is also critical for cashflow and for the distribution of resources. Since time is a critical element in the programme construction process, owners and contractors risk incurring additional and substantial costs when the construction of programmes is finished beyond the contractual completion dates set, including project milestone completion dates. In addition, programme management, a complex process that can encounter many disruptions and unexpected conditions, needs extensive networking, planning, scheduling, optimisation and decision-making tools capable of handling disputes involving completion, extension of time for different projects and informing the parties of their remedial obligations when adjudication, mediation, expert determination or, as is often the case, arbitration cases are decided. Also, many projects are to be utilised when the construction works are completed, and planning a programme can identify the date each project within a programme could be ready for operation. Therefore, a programme can have different sectional completion dates called sub-programme completion dates.

When dealing with programme management and planning, the problem becomes much more complex than the planning of a project since the critical path method and the networking methods are multi-faceted. Programmes, that comprise many projects, sometimes as many as a few hundred projects running simultaneously and in different geographical locations, need a very well organised, well thought out and complex planning method in order to construct the programme within the time specified.

Programme planning and networking involves identifying the activities that are fundamentally different to the ones implicated in project planning and scheduling, as the stages and the inter-relationships of the activities between the milestones are unique to each programme. If project planning and networking is almost micro-planning, programme planning and networking are macro, where the decision making is done on a higher level and involves activities usually done on a strategic level and includes maximising cash and financial resources; design and its stages; mobilisation, starting and completion of each project within the programme; human resources allocation; heavy equipment distribution; availability and time of delivery of critical materials; the completion of each project within the programme and the facility management if so required.

For a simple project planning, such as the planning of a building, the activities usually relate to site development, foundations and substructure activities, superstructure activities, electrical works,

Figure 6.1 Comparison of programme and project planning

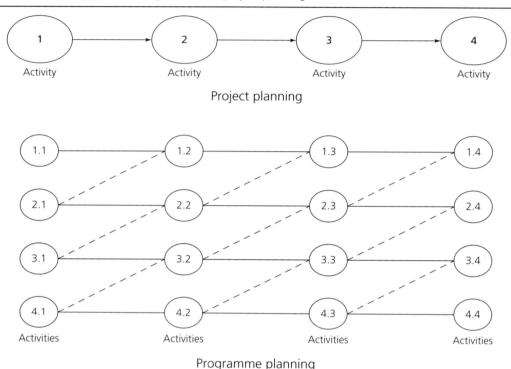

mechanical and plumbing, façade and curtain wall erection, masonry and partitions and finishing works generally (see Figure 6.1).

6.2.   Project planning – a critical path method

Engineers are often responsible for planning and supervising projects. Projects can be defined as a collection of work tasks or activities, each of which must be performed before the project is completed. Each work task entails selecting, evaluating and enumerating the resources available such as material, equipment, labour and finance necessary to ensure its completion.

Procurement at a specific time requires proper scheduling in order to have no delays in executing different activities. Planning a project involves selecting the methods to be used and determining a specific work order from all the various ways and sequences in which the job could be done. Scheduling a project requires determining the timing of the work task activities that comprise the project and coordinating them so that the overall project time can be determined. Supervising a project involves conceiving, implementing and monitoring an information system that will permit the project status to be evaluated at any time. Corrective actions are always required and initiated in order to ensure the smooth progression of the project as planned (Lewis, 2011).

In all these organisational phases the engineer is concerned with identifying components (work tasks) and structuring them into a coherent whole system. A modelling method for organisational systems associated with project planning, scheduling and supervision was introduced to the construction industry about 1960 and is known as the critical path method or CPM.

As a management tool in construction projects, the CPM scheduling technique is well known and widely accepted. In fact, many construction contracts, especially those written for large public and private projects, require contractors to submit and routinely update CPM showing critical and non-critical activities. In such projects, another occurrence is also commonplace: many participating parties often attempt to appropriate float time shown in the CPM schedules to advance their own interests.

Critical path methods require that linear graph models be formulated to specifically represent the unique features plan for the project under consideration. Organisational problems are modelled as simple graphical problems using the following steps:

(a) The various project activities are collected together and synthesised in a connected linear graph to show the project logic and sequential nature of the various activities.
(b) A variety of attributes associated with the project work tasks are added to the model.

A simple organisational problem is that concerned with the sequencing of work tasks (or activities) when each task has physical or technological prerequisites that must be met before that task can begin. Thus, sequencing activities leads to directed path network models. CPM is a graphical process requiring the development of a graph model for the project. A first step in the CPM analysis is to compile a list of activities that collectively comprise the project. The number or extent of the activities defined depends on the project, the level of detail required, and the intended management use of the model (see Figure 6.2) (Haugan, 2002).

Assigning times for completing each project activity enables the linear graph to serve as a scheduling model. Thus, it is possible to determine the following:

(a) The earliest time at which an activity may be started (known as the earliest start time or EST). EST represents the earliest time that a given activity can begin after the initiation of a project. This time is a function of the other activities that must be completed prior to starting the particular activity under consideration. The earliest start time can be ascertained by determining the maximum necessary time required for preceding activities. This requires summing the time requirements along each linear graph path from the starting point to the activity involved.
(b) The minimum time in which the total project may be completed. This is given by the duration of the maximum earliest start time path. That earliest start time path from the start of the project which determines the project completion time is known as the critical path.
(c) The latest time at which an activity may be started if the project is to be completed in a minimum time (known as the latest start time or LST). LST is specified in terms of the time from the start of the project; however it is computed backwards in time from the project finish node based on the minimum project time.
(d) The earliest time at which an activity may be finished (known as the earliest finish time or EFT).
(e) The latest time at which an activity may be finished if the project is to be completed in a minimum time (known as the latest finish time or LFT).

Figure 6.2 Typical CPM for project management

ID	Task	Duration
1	Sub-Structure:	27 days
2	Mobilise equipment	2d
3	Excavate footings	3d
4	Prepare footings	2d
5	Pour concrete footings	2d
6	Form ground floor slab	5d
7	Prepare rough plumbing	4d
8	Prepare rough electric	3d
9	Prepare rough mechanical	4d
10	Prepare and pour ground floor slab	2d
11	Super-Structure:	68 days
12	Form, prepare and pour (1st) floor col	3d
13	Form, prepare and pour (1st) floor slab	5d
14	Form, prepare and pour (2nd) floor col	3d
15	Form, prepare and (2nd) floor slab	5d
16	Form, prepare and pour (roof) col	3d
17	Form, prepare and pour roof deck slab	5d
18	Erect exterior block walls	6d
19	Install roof dry-in	3d
20	Interior wall framing	4d
21	Electric wall rough	3d
22	Plumbing and mechanical rough	2d
23	Install and finish drywall	20d
24	Interior wall finishes	12d
25	Exterior wall finishes	15d
26	Install cabinets and fixings	6d
27	Trim plumbing, electrical and mechanical	5d
28	Exterior hardscaping and landscaping	5d
29	Testing and commissioning	3d
30	Start to occupancy	95d

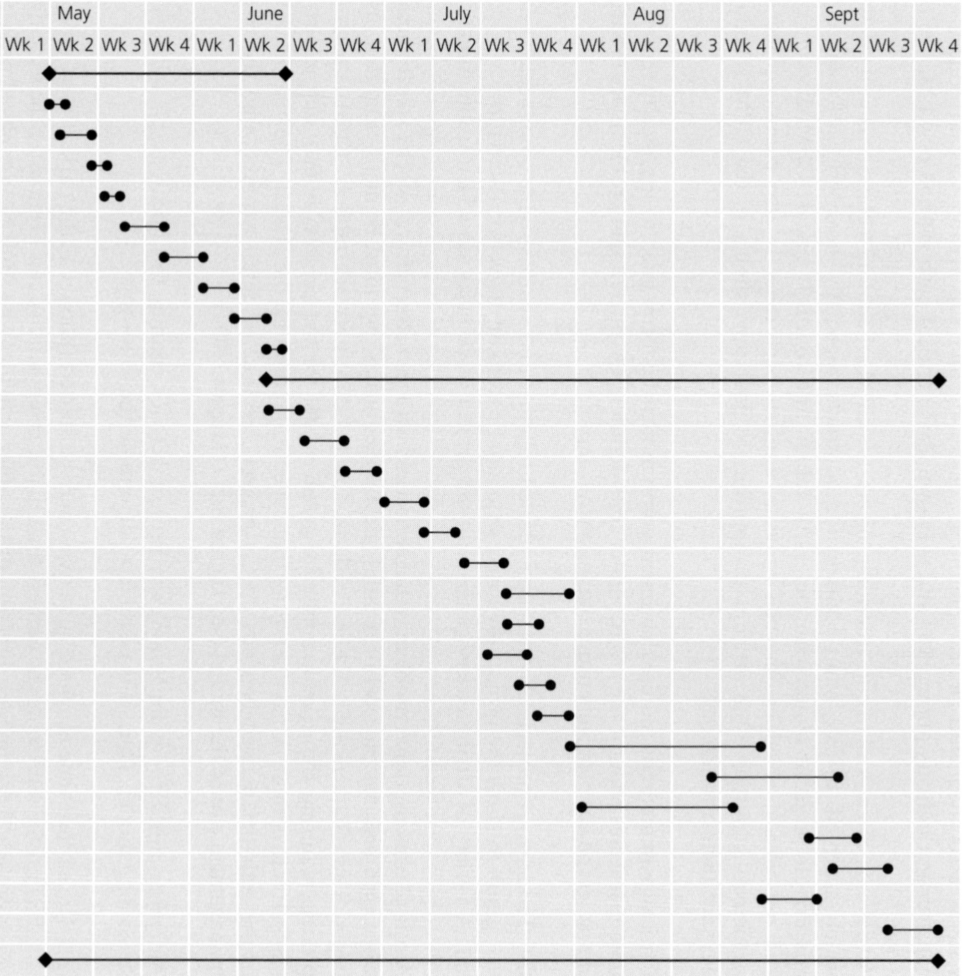

6.3.   Programme planning

For a programme, the planning and networking are completely different. Although the technique is similar, the variables and the activities are related to the programme and the projects are then considered activities of the programme.

The essential parts and activities of a programme are in sequence:

(*a*) design
(*b*) contract formation and pre-construction studies
(*c*) selection of contractor(s)
(*d*) construction phase of the projects comprising the programme
(*e*) human resources
(*f*) technical support
(*g*) supervision
(*h*) facility management.

In this chapter, we will focus on the major parts and activities that need to be tied into a generic planning and scheduling programme for programme management generally. The activities identified are major activities and the programme planner can subdivide each activity into more sub-activities depending on the level he or she wishes to explore (see Figure 6.3).

Planning the complex sequences of projects within a programme, and their dependencies, is one of the principal skills of the successful contractor or programme manager. All but the most simple of projects will proceed from such a CPM. If delay allegations are to be shown effectively by the contractor and considered properly by the architect or engineer it will be found that in most situations

Figure 6.3 Relationship between programme activities

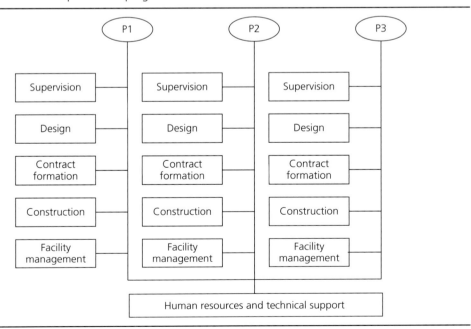

Figure 6.4 Simple CPM layout for a typical programme

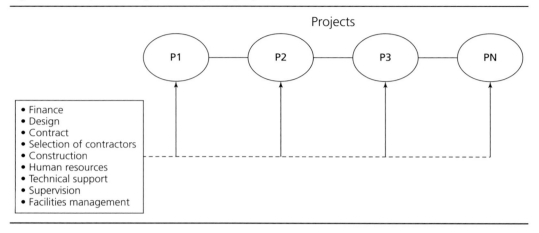

that a properly prepared CPM indicating quantity output, physical progress, as well as the passage of time is essential. The CPM facts, when married to law, must persuasively demonstrate the desired and sought-for result by virtue of the justice, equity and fairness of each party's position (Payne and Turner, 1999).

Only the largest construction programmes have traditionally been designed and built by using the CPM as a planning tool. Therefore, in minor and medium projects, when delays occur the parties involved in the contract have no structured and formal time analysis procedure for them to refer to in order to identify the quantum of the delay and the party responsible. This problem is further emphasised when more than one party is involved in the dispute and therefore the split of blame or responsibility of the delay is further complicated (see Figure 6.4).

6.3.1 Activities in a programme

The activities for a typical programme are

(*a*) financing and cash availability
(*b*) design
  (i)   appointment of the architect and design team
  (ii)   master plan
  (iii) prototype design
  (iv)   concept design
  (v)   schematic design
  (vi)   design development
(*c*) contract formation and pre-construction studies
  (i)   appointment of consultant and quantity surveying
  (ii)   composing standard contact for infrastructure and common works
  (iii) contract composition for the different projects or packages
  (iv)   finalising bill of quantity
(*d*) selection of contractor(s)
  (i)   pre-qualification of contactors
  (ii)   selection of contractors
  (iii) technical and cost submittals

    (iv) bid bonds

    (v) analysing tenders and selection of contractors

(e) construction phase of the projects comprising the programme

    (i) performance bond and bank guarantee submittal by contractors

    (ii) mobilisation of contractors

    (iii) construction phase

    (iv) testing and commissioning

    (v) completion and handover

(f) human resources

    (i) recruiting of engineers and staff

    (ii) recruiting of labour

    (iii) equipment distribution

    (iv) insurance and administration of projects

(g) technical support

    (i) planning and scheduling

    (ii) shop drawings

    (iii) quantity surveying

    (iv) invoicing

    (v) quality assurance

    (vi) quality control

    (vii) value engineering

    (viii) health, safety and environment

(h) supervision

    (i) appointment of engineer

    (ii) mobilisation of staff to different sites

(i) facility management

6.3.2   Planning and scheduling a programme

A properly working programme and schedule for a programme comprising many projects is able to address many of the problems with respect to time, delay and causation, yet there is often a reluctance by the employer or by the consultants that advise him to allow the initial tender programme the status of a contract document. This is really an essential part of planning a programme as it is often overlooked or not done properly. Usually, a programme is given a set time and then is divided into three main activities, the design period, the selection of a contractor(s), and the execution period.

This reluctance often stems from the fear that the contractor(s) is more proficient in the use of programming techniques and, therefore, is able to use the programme, usually in the form of a CPM, to his advantage and conversely to the employer's disadvantage. As a result, the opportunity to make use of the CPM to analyse post-contract time-related disputes is lost and disputes are more likely to ensue. For each programme the client must ensure that his networking analysis envelops all the aforementioned items to ensure a proper use. Another point worthy of consideration is that in a programme the client or owner is usually concerned about the completion of the programme and is not conversant on how to plan the programme. Also, the engineer, the consultant or the advisor working for the client in advising and planning the programme do not have the requisite capability, know-how or experience in dealing with programme management. More likely, they will approach the programme as a large project and try to implement the same techniques in planning and scheduling on the project and apply it to the programme, which leads always to bad advice and to misconceived ideas of how to plan the programme (Cooke and Williams, 2009).

Delays have been found to be the most cited source of disputes and the most costly cause of problems on construction programmes in many contractual regimes – programmes constituting multiple projects also face the concept of time at large, as such programmes seem to have no end in sight. Given this state of affairs, it is also noticeable that the cases that have come before the courts, where time disputes involving delay and causation issues are central to the proceedings, rarely involve the use of programming techniques as a method of reliable analysis. The traditional forms of contracts make little contractual provision to integrate the programming of activities into structural obligations for a programme. Delay and disruption issues that ought to be managed within the contract all too often become disputes that have to be decided by third parties such as adjudicators, arbitrators, judges, dispute review boards, etc. only after the delays have occurred and disputes have arisen.

6.4.   Critical path method: an analytical review

Programme scheduling consists of updating current and target schedules for existing programmes, and developing breakdown structures, milestones, target schedules and cost-loaded schedules for new programmes. The resolution of disputes on large construction and engineering contracts increasingly involves the use of delay analysis techniques to assist in the identification of the cause of critical delay to a programme and to assist in the computation of claims for lost productivity. Whilst the industry is becoming more and more familiar with the use of the tools and techniques employed in the process of delay analysis, unfortunately at present there is very little common agreement upon their correct application and understanding.

The CPM is a tool that demonstrates the shortest possible path to completion at any stage by breaking down the inter-relationship of the discrete elements that comprise the activities to be undertaken. Therefore, the CPM provides a tool by which actual job progress against a plan is monitored, thus enabling an early alert of actual and potential delays which could adversely affect the programme completion date. It is a mathematical and logic tool that can be used to predict how long it will take to complete a series of activities. A well-constructed programme using the CPM allows the parties to identify which activities, i.e. parts of the project, are critical. Any delays to the activities falling on the critical path are likely to cause delay to the completion date of the project.

A critical path analysis can only ever be regarded as a model that approximates the sequence and duration of the operations and activities on site. The perfect model would follow each resource around the site and show what it was doing and in what location and the sequence in which it carried out its work (Levy, 2012).

6.5.   Impact of the critical path method on liquidated damages

Delay in executing certain projects within a programme, which results in too many programmes finishing late and over budget, is often supplemented by enormous claims for compensation or liquidated damages. A delay can be the time during which some part of the construction programme has been extended beyond what was originally planned, due to an unanticipated circumstance or circumstances, or it can be an incident that affects the performance of a particular activity, without affecting project completion. It is possible for a delay not to extend the completion date, but nevertheless to increase the cost to complete particular activities and therefore have the potential for fuelling delay claims.

As the complexity of projects and the requirements for scheduling have increased, so have the opportunities for delay to the various activities which have been scheduled and are necessary for the completion of the programme. In fact, even determining whether completion of the total programme or a phase of the programme, consisting of a number of projects, has been delayed can be a difficult analytical task. Since delay usually leads to cost increases, there is a need to correctly determine the allocation of delay, the causes and the responsibility of the parties. With this allocation, there can be a technically sound foundation for acceptable resolution of delays' cost attribution. With the increasing use of CPM the process of sorting and recognition of the varying situations is facilitated.

In formulating and agreeing a CPM, particularly where the client is responsible for scheduling together the activities leading to the construction of a programme, a prudent contractor will normally make some allowance in the critical path for delays which might occur for which he would not be entitled to extensions of time (cashflow distribution; design stages critical to the completion; selection of the contractors according to their experience, financial stability and their pool of resources; the appointment of the consultants and/or the engineer; shortage of labour resources; late delivery of materials; or contingency time to allow for unexpected problems encountered in the execution of the works such as the appointment of specialist sub-contractors or overseas orders).

6.6.   Extension of time

The CPM can be used to minimise potential time-related claims, to justify actual claims and to assist in negotiating timely solutions of both in- and out-of-court settlements for contractors executing certain projects within a programme or in some instances executing the whole programme. In order to show that an event was not on the critical path, the defendant has to argue that the claimant's version of the critical path is incorrect and must prove on the balance of probabilities that the critical path went elsewhere. Of course it will also be appropriate to investigate whether on the facts the event actually caused a delay and whether it had any consequential effects, as well as to look for other events that may have driven progress at that time and which were therefore the true cause of delay.

There are two schools of thought or methods on how the extension of time should be calculated where an extension of time is granted during a period of culpable delay, which is a delay wholly the responsibility of the contractor.

The first method described as the 'gross' method has been preferred by many academics and some commentators and propounds that if an extension of time is granted because of an event arising during a period of culpable delay, then the extension of time must begin to run from the date the event occurred. This means that the engineer and/or consultant must establish a new completion date for the contract which adds the extension of time from the date of the instruction, thus denying the employer liquidated damages up to the new completion date. Naturally, many employers found this to be unfair.

The second method is known as the 'net' method of calculation. In these situations, it is argued that the contractor is only entitled to an extension of time equal to the time required to carry out the additional work. Effectively, this means that if the contractor is six months in delay for a project within a programme and is delayed by one further month due to a relevant event to the same project, the completion date would be extended from the original completion date to a month later, still leaving the contractor with five months of culpable delay and the threat of liquidated damages. Some contractors would consider this to be unfair as the employer may be directly responsible for the late relevant events, e.g. issuing instructions for extra work.

If there are two concurrent causes of delay, one of which is a relevant event and the other is not, then the contractor is entitled to an extension of time for the period of delay caused by the relevant event, notwithstanding the concurrent effect of the other event. Thus to take a simple example, if no work is possible on a project for a month, not only because of lack of cashflow (a relevant event), but also because the contractor has a shortage of labour (not a relevant event), and if the failure to work during that month is likely to delay the works beyond the completion date by one month, then if he considers it fair and reasonable to do so, the engineer is required to grant an extension of time of one month. The engineer is not precluded from considering the effect of other events when determining whether a relevant event is likely to cause delay to the works beyond completion. Obviously, concurrent delays become much more complex when dealing with a programme and concurrent delays affect different projects within the programme.

Liquidated damages are damages calculated according to a formula set out in the contract for each project or for the whole programme. They are designed to save the employer from having to prove the actual damage suffered. A clause in a construction contract providing for liquidated damages for delay has to be closely linked with a clause which provides for an extension of time. This is because it is to be assumed that if the contractor is to be held liable for liquidated damages, the delay for which damages are to be calculated must be the responsibility of the contractor. If delays caused are the responsibility of the employer then the contract must provide a mechanism whereby the date for completion can be extended.

If the contract fails to provide for such an extension, or if the architect or engineer fails to administer the extension of time provisions correctly, then the liquidated damages clause may become unworkable, if only because there is no fixed date from which to calculate the delay for which the contractor is responsible.

In order to prove a delay claim it is necessary for the claimant to satisfy the requirements of the particular contract under which he is operating. This is likely to mean first that a particular event is established as being of a type that gives rise to an entitlement under the contract for an extension of time, second that the event has occurred and its extent is quantified, and third that its consequences are identified by establishing that the event caused delay to the completion of the project. In the context of construction projects, however, the latter is often very difficult to do, and the extent to which this is possible is usually as much to do with the quantity and quality of the records held by the contractor as to do with what actually happened (see Figure 6.5).

6.7.   Planning complexity and float

The more activities there are in a programme, the greater will be the number of logic links between them and the greater the number of assumptions involved in completing the model. Hence, when carrying out retrospective delay analysis using Critical Path Analysis, large CPMs with hundreds and thousands of activities will produce unreliable results. This is because the analyst has made hundreds of assumptions in preparing the CPM and, when considering the impact of an event, he would be likely to make many adjustments to a programme if faced with a potential delay, especially if the delay is on the critical path.

It is important to recognise that it is easy to manipulate a CPM for a programme in order to derive the required end result. For example, if a planner wishes to make a certain number of projects within a programme critical, he achieves this by fixing durations of each project within the programme or logic links between the completion dates of the projects. Equally, if there have been

Figure 6.5 Extension of time methods

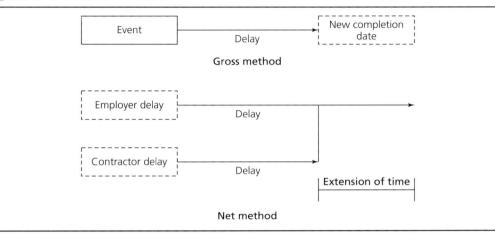

Gross method

Net method

variations issued in one part of the works, it is possible to make this element of the programme critical. This is another reason why it is worth considering reducing the number of activities on very large programmes used in retrospective delay analysis (Lester, 2007).

The float ownership concept is fundamental to the analysis of programme delay and the allocation of responsibility. Both owner and contractor want access to the float in the schedule because it affords them more flexibility in their decision making and use of resources. However, many tailor-made programme contracts do not address this important topic. As a result, neither the owner nor the contractor has a contractual right to use the float. The now generally accepted, and sometimes disputed, answer is that the programme owns the float with its own complexity. Under this interpretation, a party is permitted to delay a project within a programme with positive total float provided that the delay duration does not exceed the total float calculation for that activity, and their use of positive total float occurred prior to anyone else's. The float ownership concept becomes more complex when the programme is late and the total float calculation becomes negative.

Contract formation should be diligently done where there is no reason why the contractor cannot show on its CPM a date for planned completion earlier than the completion date, thus including some terminal float in its schedule and, hence, a delay to the completion date is assessed as the length of time so that, due to the compensation event, planned completion is later than planned completion as shown on the accepted programme. Therefore, as should be stated in the guidance notes to the contract, any terminal float resulting from an early planned completion date of a programme is preserved. The period of delay to the planned programme is then added to the completion date to determine the revised completion date from which delay damages will be applicable (see Figure 6.6).

6.8. Acceleration of a programme

An employer may choose to reverse the impact of delays by expressly ordering acceleration in order to put the programme back on schedule and applying CPM analysis showing the cost of such a directive. In other cases, where the owner resists granting the contractor an extension of time that he is entitled to for an excusable event or where an extension has been granted by the

Figure 6.6 Float for a programme

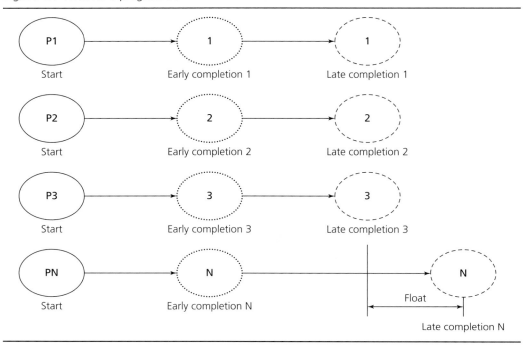

owner, but for a shorter period than the contractor is entitled to, a contractor may feel compelled to accelerate the works for a certain project within a programme in order to overrun the completion date set by the owner, thereby avoiding exposure to liquidated damages. To recover under this theory the contractor must prove:

(a) The owner failed to grant an adequate or any extension of time for a project.
(b) The owner did not make it clear that completion was required within the original contract period.
(c) Adequate notice had been given by the contractor to the owner advising that he was treating the owner's actions as constructive acceleration.
(d) The proof that there had been the actual insurance of additional costs.

If the programme contract provides for acceleration, payment should be based on the terms of the contract even if such acceleration is directed, and directed to parts of the programmes as each project or part of a programme is assessed as a separate entity. If the contract makes no provision, the parties should agree the basis of payment before acceleration is commenced.

6.9. Risk distribution in managing delays

It is a generally accepted principle of risk management that those who are most able to manage a particular risk should bear that risk. However, in construction and engineering standard forms of contracts, that principle becomes absolute. Once the contractor is appointed and the work commences on a programme or part of a programme, all the risk of change or interference that is left in the hands of the employer cannot be managed because all the information needed to manage that risk is left in the hands of the contractors. Of course, the contractor(s) is required to use the tools he has to manage the employer's risk and to overcome and avoid unnecessary delay howsoever caused. However, if he fails to do so, he is to be compensated for any loss he suffers and is to be given more time to complete. History has shown the possibility of the contractor managing the employer's risks to be non-existent and it is perhaps not surprising that under the current standard forms of contract there is little impetus for the contractor to do so. The main parties involved in the contract are often reluctant to allow the initial tender programme to have the status of a contract document. This reluctance often stems from the traditional hierarchy of the construction industry, and the in-built fear that the parties involved lack the expertise in the programme techniques and furthermore the practical application of such programme techniques in order to achieve the target time and eventually cost. Because of the nature of construction, the standard forms of construction contracts tend to have a certain amount of flexibility built in that enable the work to be carried out to be varied as it proceeds, thus enabling both parties to fulfil their obligations in accordance with the terms to which they are contracting.

In terms of the CPM use during the post-contract period of a programme, although the CPM will provide a support in terms of better understanding on the impact of delays it will not necessarily establish the extent of delay unless more sophisticated forms of analysis are adopted and are used on an ongoing basis throughout the project so that the CPM becomes literally a working and living CPM. This is because if delay allegations are to be shown effectively by the contractor(s) and considered properly by the consultant or engineer, it will be found in most situations that a simple bar chart will not suffice and some better means of indicating quantity output or physical progress, as well as the passage of time, is essential. The purpose of CPM in connection with claims is to identify the causes of delays, the dates of onset and cessation and also, as far as possible, the effects both immediate and consequential on the various operations.

The emergence of new Standard Forms of Contracts, in particular the New Engineering Contract, where planning and time are given a more dynamic status, and the gradual recognition that the planner should be an integral member of the employer's design team have, therefore, provided the employer with the ability and confidence to accept the tender CPM as a contract document for the whole programme. Parties to a programme construction contract often end up in disputes because the form of contract under which the programme or part of the programme comprising a number of projects is procured whilst providing a mechanism for dealing with delay and disruption does not guide the parties in terms of how the resulting delays should be prepared and managed, and what procedures should be followed during the currency of a contract to enable solutions to be arrived at that sufficiently prevent the parties from falling into dispute.

In terms of minimising associated risks, it is critically important that the CPM should be updated when executing a programme. In fact there should be a contractual requirement for the accepted CPM to be updated with actual progress at regular intervals during the currency of the contract and the interrelationship of each project with the rest of the projects within a programme. The aim is that the CPM becomes a proper management tool that enables the monitoring of actual progress against planned progress.

6.10. Completion, early completion and acceleration

If as a result of an employer delay, the contractor is prevented from completing the programme or parts of a programme by the contractor's planned completion date (being a date earlier than the contract completion date), the contractor should, in principle, be entitled to be paid the costs directly caused by the employer's delay, notwithstanding that the employer is aware of the contractor's intention to complete the works prior to the contract completion date, and that that intention is realistic and achievable.

If, as a result of a compensable employer risk event that causes delay, the contractor is prevented from completing the works by his own planned completion date (being a date earlier than the contract completion date), the contractor should, in principle, be entitled to be paid the costs associated with such an event. In other words the float associated with the contractor's planned completion date belongs to him. The date for completion becomes the date by which the contractor expects to complete the works. This confirms the view that in order to establish that an event has affected the completion date you must show that it falls on the critical path. Some means by which the critical path can be identified and shown therefore appears to be required. Although the emphasis appears to be on an examination of what actually happened, the passage does not make it clear whether the critical path at the time of the event or the overall as-built critical path evident at the completion of the programme.

In terms of the completion date, the planner is more specific than some of the more traditional standard forms, such as those under the JCT umbrella. A contractor under a JCT form is entitled to plan to finish the works before the completion date stated in the contract, but the fact that he has done so does not place any obligation on the engineer to produce information by dates earlier than would be necessary for completion by the contract date. It is an established fact that an employer does not have to assist the contractor in his efforts to complete early, but nevertheless he should refrain from doing anything that would deliberately hinder early completion.

6.11.   Programme float ownership

Float is the amount of time by which an activity or group of activities may be shifted in time without causing delay to a contract completion date of a series of projects or a programme as a whole. The ownership of the float, which may ultimately determine entitlement to an extension of time as a consequence of employer delay, should be adequately addressed in the contract. A contractor should not be automatically entitled to an extension of time merely because an employer's delay to progress takes away the contractor's float for a particular project or a major activity leading to the delay of a project within a programme. An employer's delay should only result in an extension of time if it is predicted to reduce the total float of the activity paths affected by the delay to below zero.

The authors distinguish float as it relates to time and float as it relates to compensation, whereas the contractor, traditionally, takes the view that float belongs to him, and that he can use it as he sees fit. This book stipulates that float is a project resource, to be used when the project needs it to achieve the programme's planned completion. In order to determine the ownership of float, the following steps should be taken:

(a) Determine what projects are affected within the construction of a programme.
(b) Calculate the event duration from all affected activities by reference to the last updating.
(c) Determine the status of the projects that are affected at the time the variation is issued or when the delay occurs.
(d) Create a detailed analysis of the sequence of projects or main activities, such as procurement, finance or design, that are necessary to satisfy the change requirements or that identify the delay.

The contractor needs a time buffer for its own use; this should be included as a time contingency in the relevant construction industry. Float is not usually dealt with in standard forms of contract, and the authors sensibly aim to overcome this. It has been suggested that this recommendation could lead to avoidance of disputes in circumstances where a principal uses the whole of the float and the contractor becomes liable for liquidated damages at the end of the programme.

In a resource-constrained schedule the concept of float breaks down and, quite often, the concept of a critical path breaks down. Since almost all construction projects are resource-constrained, at least to some extent, this becomes a source of major problems. The classic legal question in recent years as well as the subject of numerous professional papers relates to the ownership of the float. It is difficult to claim ownership of something that may not exist or that has not been quantified properly.

The resolution of disputes on large construction and engineering programmes increasingly involves the use of computer-based delay analysis techniques to assist in the identification of the cause of critical delay to a project and, in the more sophisticated cases, to assist in the computation of claims for lost productivity. Whilst the industry is becoming more and more familiar with the use of the tools and techniques employed in the process of delay analysis, the future will bring more common agreement upon their correct application. The problems of unresolved delay and disruption in construction contracts are notorious. Unintended delay causes disputes and losses in construction and civil engineering contracts worldwide and the common view in the industry is that many disputes arise because the parties do not understand the way delays occur and how their consequences could be avoided.

6.12. Programme structure within a programme

Structure may be considered as the established pattern of relationships among the components or parts of the programme. We agree that the structure of the programme management of a programme cannot be looked at as completely separate from its functions; however, these are two separate phenomena in which when they are viewed together, the concepts of structure and process can be seen as the static and dynamic features of the organisation of a programme. In some decision-making techniques for programme managers the static aspects (the structure) are the most important for investigation. These are the projects and the components of the programme to be built. In others the dynamic aspects (the processes) are more important. These consist of the factors assisting in building the programme such as finances, resources and engineers and labours (Nieminen and Lehtonen, 2008).

The formal structure is frequently defined in terms of the following:

(a) the pattern of formal relationships and duties – the programme management chart plus job descriptions or position guides
(b) the way in which the various activities or tasks are assigned to different departments and/ or people in the programme management (differentiation)
(c) the way in which these separate activities or tasks are coordinated (integration)
(d) the power, status and hierarchical relationships within the programme management (upper management's decision making for programme management)
(e) the planned and formalised policies and controls that guide the activities of staff in the programme management team.

Formal organisation within a programme is the planned structure and represents the deliberate attempt to establish patterned relationships among components that will meet the objectives effectively. The objectives are always to complete the construction of the programme within the time, budget, specifications and constraints. To reiterate a comparison made in Chapter 1, if project management has two dimensions then programme management has three dimensions.

The informal programme management refers to those aspects of the decision making for programme management that are not planned explicitly but arise spontaneously out of the activities and interactions of participants. Informal programme managements are vital for the effective functioning of the programme management.

It is impossible to understand the nature of a formal programme management within the hierarchy of a programme without investigating the networks of informal relations and the unofficial norms as well as the formal hierarchy of authority and the official body of rules, since the formally instituted and the informally emerging patterns are inextricably intertwined. The distinction between the formal and the informal aspects of programme management life of a programme is only an analytical one and should not be reified as there is only one actual programme management.

The concept of programme management planning implies the process of developing the relationship and creating the structure to accomplish management and programme management purposes. Structure is, therefore, the result of the planning process. Programme management structure has a perspective and action orientation; it is geared to solving problems and improving performance. Managers are the primary designers of most programme management. Although they may have help from lawmakers, consultants, researchers and academics, managers ultimately must make the major decisions.

Programme management planning is never complete; it is a continuing, ongoing process. A well-designed programme management plan and schedule is not a stable solution to achieve, but a developmental process to keep active.

Coordination of activities is an important consideration in the planning of programme management structures. Integration is defined as the process of achieving unity of effort among the various sub-decision-making entities for the accomplishment of the programme manager task. The requirements of the environment and the technical decision making for programme management often determine the degree of coordination required. In some programmes, it is possible to separate activities in such a way as to minimise these requirements.

6.13. Responsibilities and function within a structure

Structure is directly related to the assignment of responsibility and accountability to various programme management units. Delegation is fundamental in the assignment of both authority and responsibility. Control decision making for programmes is based on the delegation of responsibility. Most programme management organisations develop some means to determine the effectiveness and efficiency of the performance of these assigned functions and create control processes to ensure that these responsibilities are carried out.

Traditional management theorists were primarily concerned with the design of efficient optimisation techniques. They emphasised such concepts as objectivity, impersonality and structural form. The programme management structure was designed for the most efficient allocation and coordination of activities. The positions in the structure, not in the people, had the authority and responsibility for getting tasks accomplished.

Authority is the right to invoke compliance by subordinates on the basis of formal position and control over rewards and sanctions. Authority and responsibility should be directly linked; that is, if a project manager is responsible for carrying out an activity or a project, he should also be given the necessary authority. Accountability is associated with the flow of authority and responsibility and is the obligation of the subordinate to carry out his responsibility and to exercise authority in terms of the established policies. This view of authority, responsibility and accountability provides the framework for much of traditional programme management theory.

REFERENCES

Cooke B and Williams P (2009) *Construction Planning, Programming and Control.* Wiley Blackwell, Oxford.

Haugan T (2002) *Project Planning and Scheduling. Management Concepts*, VA.

Lester A (2007) *Project Management, Planning, and Control, 5th edn.* Elsevier, Oxford.

Levy S (2012) *Project Management in Construction, 6th edn.* McGraw-Hill, New York, NY, USA.

Lewis J (2011) *Project Planning, Scheduling, and Control: The Ultimate Hands-On Guide to Bringing Projects in On Time and On Budget, 5th edn.* McGraw Hill, New York, NY, USA.

Nieminen A and Lehtonen M (2008) Organisational control in programme teams: an empirical study in change programme context. *International Journal of Project Management*, **26(1)**: 63–72, **doi**: 10.1016/j.ijproman.2007.08.001.

Payne JH and Turner JR (1999) Company-wide project management: the planning and control of programmes of projects of different type. *International Journal of Project Management*, **17(1)**: 55–59, **doi**: 10.1016/s0263-7863(98)00005-2.

Programme Management in Construction
ISBN 978-0-7277-6014-2

ICE Publishing: All rights reserved
http://dx.doi.org/10.1680/pmic.60142.113

Chapter 7
Design in programme management

7.1. Design definition

Design is the creative and disciplined action of problem solving, both individually and in teams, in order to deliver real, useful, elegant and enduring outcomes in the built and natural environments.

Renaissance architect and philosopher Leon Baptiste Alberti identified the tenets of good architecture as 'firmness, commodity and delight'; today, only 'sustainability' need perhaps be added.

Design may be defined as the realisation of a concept or idea into a configuration, drawing, model, mould, pattern, plan or specification (on which the actual or commercial production of an item is based) and which helps achieve the item's designated objective(s).

7.1.1 Programme management definition for design

Programme management can be defined as the organisation of multiple, complex problems and information into a logical, disciplined sequence of related, inter-related and unrelated actions, which can then be sorted into solutions of meaningful actions by clients, consultants and contractors, leading to programme delivery and individual project delivery within the programmes, on time, to budget and at the right quality.

7.1.2 Good design principles

In programme management, good design teams strive together to embrace, understand and creatively apply all aspects of complex technical knowledge and new emerging technologies to the individual projects and overall programme.

Good designers working collaboratively in multidisciplinary teams can help the programme manager to achieve this overall objective of sound management, both elegantly and efficiently but developing a well-structured knowledge base tree.

The evolving design knowledge base is held collectively by the design team, controlled by the design team leader for decision making by the programme manager working in collaboration from the planning stages through to completion (see Figure 7.1).

A design knowledge base is ideally shared parametrically in building information management models (BIMM) with the wider programme teams including construction teams and with the full understanding of the client–owner team, in real time, to track, evaluate and integrate new information as it is generated and then submitted to the programme director in sequence, in easily digestible graphic form. It can then be evaluated, integrated and approved for construction with confidence in true delivery of appropriate cost, quality and time.

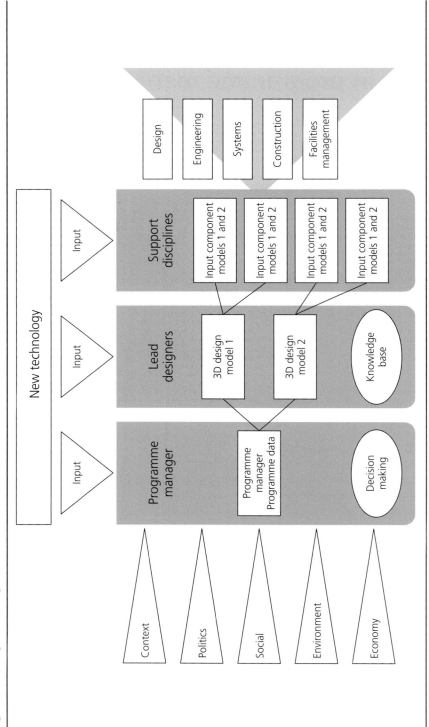

Figure 7.1 Design knowledge base

Good designers and their support disciplines strive to keep track of broad new developments and new technologies in the broad design, engineering, systems, construction and finally facility management (FM) industries.

Current information and real-time dating of the knowledge base have proved essential in major international programmes. The programme manager also needs to track and integrate local changes in the international, national and regional contexts.

Physical, political, social, environmental and commercial inputs combined with specific developing human factors, all affect the overall programmes, as it develops with the subsidiary individual projects.

However, good design can be paradoxical and difficult to define, as detailed below.

The essence of good design in programme management is inspired programme and design leadership, working collaboratively with all the necessary talented team members and all required resources, focused on common objectives within an appropriate time frame: a fully integrated design action.

7.1.3 Integrated design approach

Good design in programme management as described above essentially represents an integrated design approach. Integrated design has been developed by many consultants in some detail. Their approach culminates in a comprehensive, collaborative professional service (see Figure 7.2).

Figure 7.2 Integrated design timeline comparison to other deliveries

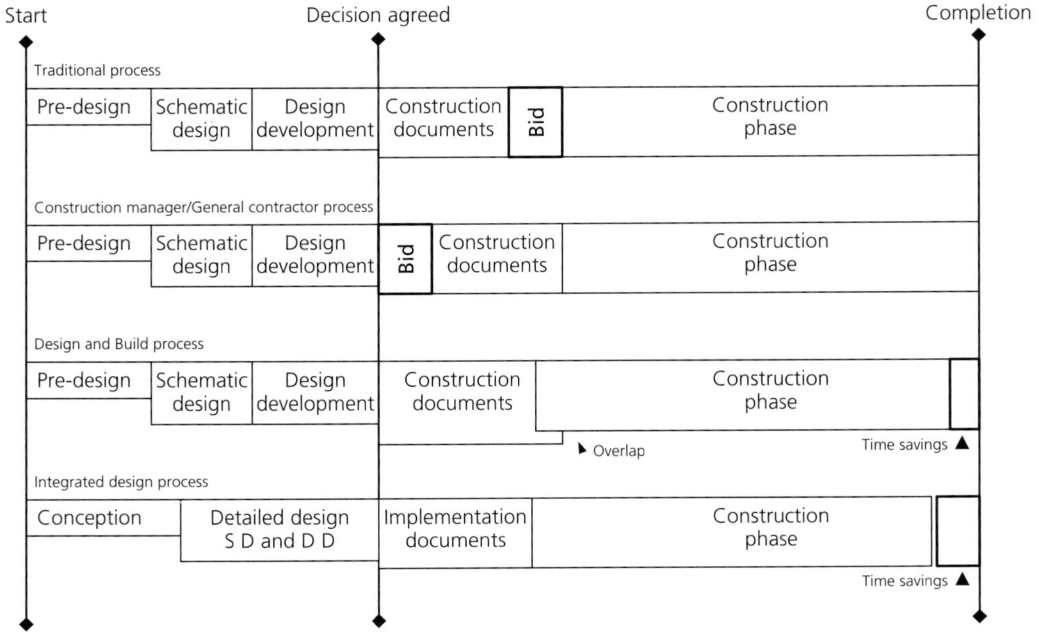

Integrated design results in considerable time and cost savings and can avoid many critical delays and causes, when compared with more outdated delivery methods.

The American Institute of Architects (AIA) has defined integrated design as:

> An ambitious, fully integrated approach to good design can only be achieved by the designers, the programme manager, the contractors, facility managers and the client with end user input, all committing to a genuine knowledge sharing endeavor, fed by regular research of publications and input from current news media from the outset.

(AIA, 2014)

7.1.4   Good design – paradox

Good design is central to good, well-organised projects and programme management but it can be an area of professional expertise that is often misunderstood and misrepresented, with potentially very serious consequences to the overall programme.

Good design thrives under **open** client and programme management as opposed to **closed** author- itarian principles, which stifle real solutions in the authors' experience.

Good design, even with sound economic innovation, is paradoxically not always the most efficient design; powerful human factors of social history, memory and sentiment all have influence on good design outcomes.

'What is good design?' is a seemingly simple question that is surprisingly difficult to answer. The more you think about it, the more complex the question becomes. Not only does 'good design' mean different things to different people, it also changes at different times and in different contexts. Some of the world's leading designers were challenged to define what 'good design' means now in a debate (at the World Economic Forum Annual Meeting, Davos-Klosters, Switzerland, 23–27 January 2013). Each designer was asked to identify one example of 'good' and one example of 'bad' design, and to explain the reasons for these choices. Typically, the issue was one of emotional appeal set against efficiency.

A similar 'emotions or efficiency' debate erupted over the design of London's buses. Londoners still reminisce fondly about the Routemaster double-decker bus, which was withdrawn from service in 2005. Popular though the Routemaster was, getting on and off it was very difficult for small children, people with mobility impairments, or people with pushchairs, and virtually impossible for wheelchair users. Those problems were solved by the design of one of its successors, the articu- lated or 'bendy' bus, which is easily accessible but which was widely loathed by Londoners (and London cyclists in particular, for its deceptive and dangerous wheel tracking in traffic).

The question therefore is which of the buses constituted 'good design'? Whilst the bendy bus was very easy to get on to and could carry twice as many passengers and more people could sit down, the old Routemaster double-decker bus looked great but was very hard for many people to use.

The rational answer would, of course, be the bendy bus, but can something be 'good design' if no one enjoys using it?

Because of the discontent about the bendy bus, a new neo-Routemaster with solar panels was developed and is being used on the London streets (see Figure 7.3).

Figure 7.3 Routemasters new and old

(a)   (b)

The Mayor of London was right to insist that the new bus be environmentally responsible, because that is another non-negotiable component of 'good design', albeit one that has attained the status relatively recently. As the environmental crisis has deepened, it has become impossible to ignore the ecological implications of everything we consume.

However gorgeous, witty, ingenious and even useful something is, we can no longer consider it to be a design success unless it is also ethically and environmentally responsible.

It therefore often remains difficult to succinctly define good design, as opposed to neatly defined engineering, scientific or empirical design and statistical quality assurance or risk management.

This difficulty in defining good design is perhaps largely due to the intuitive and creative aspects of the traditional design process combined with the computer-aided and newer 'parametrically powered' design process, allowing designers to make ever more rapid leaps of faith and imagination with increasing levels of three-dimensional quality, accuracy, constructability, sustainability, time, float and delay management and consequently vastly superior cost prediction (see Figure 7.2). Risk prediction and management using Monte Carlo analysis and fuzzy logic are integrated logically with parametric design.

It is therefore precisely this paradoxical quality of design, combined with new technology, which can permit the well-rounded programme manager to prevail in his role of (to expand upon the definition coined by Ali D. Haidar in Chapter 3 of this book) diagnostician, pragmatist, philosopher and artist.

For this critical reason, the design role in programme management is described in some detail below, to begin to illustrate general professional best practice principles, rather than to apply rigid but rather scalar guidelines for programme managers striving for excellence.

Designers are able to move forward from well-established past experience and precedent, onward to exciting new, programme- and project-specific solutions, which satisfy much more than the narrow criteria confines of simple 'efficiency'. They do this creatively and intuitively, by reaching into areas of deep human factors of satisfaction, spirituality and delight, backed up by new tracking and accounting capabilities of BIMM, algorithmic design and computer-based artificial intelligence networks and systems which include constructability, risk management and quality assurance.

This 'human factors friendly' yet more predictable aspect of good design is particularly relevant and visible in urban place-making, where human scale, natural security surveillance, use of noble materials to create harmony and beauty of landscape, lighting, water and enclosure are visible and used publicly.

In the hands of a master designer, these elements can all work together to create recognised and well-loved design iconical and analogical design solutions in many different contexts: landscaped public spaces; grand civic architecture; vehicles; interiors; furniture; industrial, commercial and humble household products; jewellery; and fine art.

To understand and to begin to demystify the design process and its relevance to programme management, design activities can be defined and broken down into certain basic traditional work stage elements with more complex over-layers to utilise new technologies.

7.2.   Design roles and responsibilities

However simple the stages become, a very sound understanding by the programme manager of the specific roles and responsibilities of all members of the design team is required for programme managers to control good programme outcomes.

Throughout the design process, identification of active roles and responsibilities remains a crucial requirement at the inception of the programme, if the overall programme objectives are to be achieved.

Confusion of roles and unrealistic expectation of responsibilities, without contractual or legal power to act to correct them, will frustrate all parties involved.

Where the designs of multiple projects each impact on one other, the roles and responsibilities each player carries should be carefully defined, regularly updated with documented change management and widely communicated in order to be crystal clear from the start.

As any changes and adjustments in staffing or job descriptions occur these should be recorded in writing and remain defined throughout.

Change control, in tracking new client instructions to the programme director or manager, should include this vital aspect of defined roles and responsibilities.

7.2.1   Traditional UK design roles

Traditionally, in the UK, professional design roles were first centred around the building surveyor, then on the architect role, first licensed in the mid-1880s, which carried overall responsibility for the design, often in collaboration with skilled artisans and craftsmen, using very limited building information.

A decade later, as building technology and changing economic conditions became more complex, particularly in terms of rising inflation for the first time in history, the architect personally employed any sub-consultants he required, such as quantity surveyors or structural and/or mechanical engineers, to assist him in keeping pace with his rapidly increasing workload. Construction techniques and activities were increasingly industrialised, which often led to product experimentation on site and frequent material failures.

The client was routinely left out of the decision-making process.

So when all these factors of building material inflation, increasing complexity of mechanical and electrical systems and market demand for the limited numbers of architects available to clients reached a climax in the 1960s and costs on design and construction soared unpredictably, clients became increasingly litigious.

The profession of project management was born.

When building design fees and construction costs spiralled out of control, the new profession of project management appeared, fresh from the event management and aerospace industries. The project manager was someone who could effectively step in between the architect and the client.

This was ostensibly to manage the design process without actively being fully involved in the design itself.

Project managers were often from quantity surveying and building cost planning backgrounds and were employed to 'rein in' architects and designers, who were seen as being out of control of the hard economic aspects of their increasingly complex projects.

Active design management was not often apparent and the project manager was seen initially by designers as a 'post box' and as having a 'penalty' input only. There was, figuratively speaking, a civil war between the professional team members, contractors and clients on many projects, resulting in the dismal quality, soulless buildings and environments left standing from this period as sad monuments today.

This unacceptable situation was partly resolved by the innovation of design and build contracts, which started to bring the parties back to realistic collaboration. However the design quality for these contracts tends to be generally lower than that of contractual arrangements where designers and constructors retain more autonomy. The solution lay in far more balanced, standard industry contract types, tailored to clearly define project scale and scope more clearly and appropriately. Project managers became more involved in the detailed time, quality and cost planning of projects, and they are now routinely drawn from all support disciplines.

Under the recent rise of BIM design techniques, where economic and other empirical factors are automatically and accurately added as attributes to 3D design modelling, the roles are changing as a response to new technology.

Programme management is by extension a product of the information age, a recent development of project management and its economic roots, to manage multiple projects of greater complexity, and must take a far more active, sustainable and collaborative approach to managing the design team.

7.2.2 Traditional US professional roles and responsibilities

In the USA design and technical documentation roles were split, in the early nineteenth century, into clearly defined collaborative but separate roles, with creative and intuitive concept sketches, design development drawings, visuals, perspectives and design models being made by architects, interior designers and landscapers on the one hand and tightly defined, construction industry-related, 2D and 3D design drawings, technical drawings, specifications, structural, mechanical and electrical (M&E) and building management services and detailed component schedules by non-professional technical drawing firms on the other hand.

This division of labour and skills had and still has many benefits in making use of professionals' differing areas of knowledge and aptitude in respect of creativity and technical knowhow but is often a stumbling block today for larger programmes as it adds another interaction and layer of potential confusion, easily eliminated by a more integrated design approach, with 'star or signature' architects zealously guarding the entire design process, aided by BIMM.

7.2.3 Supporting discipline roles

As the lead designers work on a complex programme, progressing steadily through the programme's stages, essential inputs from core team support disciplines and specialists are needed to develop and progress the design simultaneously.

Examples of typical core team support discipline professionals are:

- land surveyors
- cost planning and quantity surveying
- civil and structural engineering
- electrical and mechanical engineering
- telecommunications and systems engineers.

Examples of typical specialist team support discipline professionals are:

- contract, legal and commercial specialists
- geographical information systems (GIS) specialists.

Core team inputs are made possible by semi-automatically generated and co-ordinated inputs of project-specific attributes using powerful parametric computer program engines, such as Autodesk Revit and to a lesser extent City CAD (computer-aided drafting).

Automation of core discipline attribute inputs, taken from architects' and urban designers' first 3D models, means that the role of architect and urban designer as the orchestrator or conductor of good design is currently rising rapidly in importance as one of the leading activities in project and programme management today.

In this sense the 'engineering approach' to design of past decades is rapidly being transformed back into a more hands-on and creative/intuitive, collaborative, integrated good design response, which is being constantly informed and subjected to semi-automatic review from all analytical disciplines, as it progresses and is adjusted in real time in BIMM. No attempt to minimise the value of fully integrated multi-disciplinary engineering teamwork is intended here.

The new emphasis on the necessary lead design role of architect or urban designer in programme management remains utterly dependent on good core team inputs, but these are simply better coordinated by the lead designer and the BIMM system.

In the rail industry the term 'lead design engineer' is often used to describe this role. This is perhaps a more accurate and stable term than 'architect', which, though a traditional role, is becoming increasingly IT-centred and consequently is changing rapidly.

A well-organised, dedicated web-based document and collaboration platform allows for all the support disciplines to make invaluable and critical contributions, with greater professional time savings, less confusion and delay in CPA and economy of construction work. This work method has been proved highly beneficial in minimising debilitating delays caused by site clashes, as fully integrated 3D parametric models are developed and realised into actual built form.

Another major benefit of BIM is the reduction of a large portion of the time and budget spent obtaining programme and project approvals. This also applies to obtaining client approvals.

Time and cost savings occur when each discipline shares its work through 3D modelling and specifications with the local authorities, and receives early guidance and comments from those officials responsible for regulating the following approvals:

- environmental impact
- regional and town planning
- transport and traffic
- building control
- civil defence
- waste disposal and
- fire and general safety.

Programme management is a rapidly growing new discipline with specific imperatives where regular contact with as well as control and coordination of all professional inputs is crucial to successful project and programme outcomes. A key emerging tool to achieve this is the charter agreement.

7.2.4 Charter agreements

A legally informal but well-written and signed charter document, including all team players in this endeavour as taking part in a broad-based partnership, can often focus the required attention on good design to help maintain and align important relationships throughout the projects and programmes, moving on to delivery and lifecycle in use.

The good charter recognises and records the fact that programme management is not a structural or stable solution to achieve but rather a developmental process action.

Formal and informal relationships are recognised and defined by the charter to analyse and set down in writing a dynamic, human resource-centred process of developing relationships creatively for the solutions needed to actually accomplish specific, international programme objectives in a locally acceptable manner. This is in contrast to strictly formal legal contracts, which tend to be tied to a specific legal national system, often irrelevant and remote from real delivery and local human and physical resource capability.

A charter is particularly useful in defining fair, equitable and just outcomes for all parties in sharing risks realistically in proportion not only to team remuneration and owner/client returns but also to end-user programme satisfaction. The latter is highly relevant in major programmes with enormous social, political and human development consequences.

A good charter will contain a broad summary of the more complex project and programme issues, to ensure a broad general understanding in the entire team of formal contract structures in place, explaining it in much more simple terms. A charter may contain the following elements:

- background history
- common understanding
- scope statement
- business case
- spending approvals procedure
- change control procedure: float ownership variables
- key milestones agreement: float ownership tracking
- outline communication plan: team
- stakeholders' communication matrix
- programme and project team house rules
- strengths, weaknesses, opportunities, threats (SWOT)
- risk register
- lessons learned record: feedback during and after the programme's life
- signature and commitment.

Good charter definition is particularly relevant on complex or urban-scale projects and programmes where the individual team members and leaders can easily lose contact with rapidly changing programme dynamics.

The London 2012 Olympics programme had 120 principal contractors working on 400 projects with 30 000 early warning and compensations events with few reported disputes and was 18 months ahead of schedule in the closing stages.

7.3. Design work stages

Once programme management roles and responsibilities are clearly defined, and financial constraints/resources are clear, design roles and responsibilities can then be more readily organised into time schedules and located in a typical work-breakdown structure (WBS), which in turn can be fed into the critical path analysis (CPA) or critical path method (CPM) planning of individual design projects and ultimately into multiple CPA streams interlinked with beneficial control of float time, delays, quality and cost in multiple project programmes.

However, to understand the individual parts of a design, work breakdown structure (WBS), design work stages and their evolution over time need to be well understood. Section 7.3.1 enumerates the stages of the design process for traditional procurement models, which are routinely adjusted in practice for variations in procurement such as design and build contracts or newer Joint Construction Tribunal (JCT) and recent New Engineering Contract (NEC) formats.

7.3.1 Seven broad stages of the design process

The design process refines programme information from the generally known to the specifically useful through seven broad stages of creative, disciplined actions:

(*a*) identifying and collecting the necessary data
(*b*) analysing and organising design data
(*c*) understanding the problem, each project objective and the overall programme objectives
(*d*) agreeing the brief, budget, time and design criteria
(*e*) proposing and communicating design options
(*f*) agreeing the preferred solution and obtaining approvals
(*g*) procuring, implementing, modifying and completing, with feedback into design.

7.3.2 Design stages by professional international design entities

The broad stages identified in Figure 7.4 are formally arranged by the various professional institutes such as the Royal Institute of British Architects (RIBA) or AIA into very detailed work stages or 'plans of work', which are fully in use in the UK, Western Europe, ECU and USA and, to

Figure 7.4 Seven broad stages of the design process

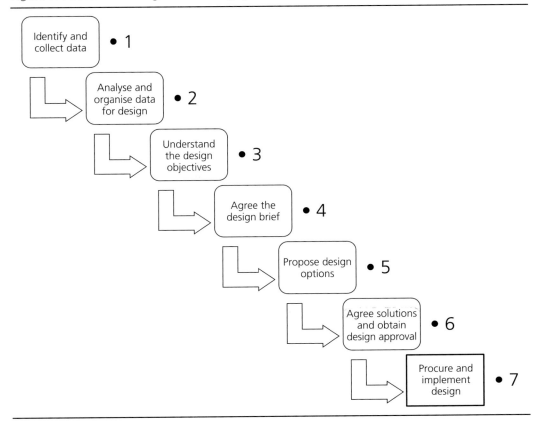

varying degrees, internationally. These are updated regularly in line with industry contracts and developments thereof, such as NBC or FIDIC, or with innovations such as BIM, and well-defined health and safety, quality assurance and sustainability overlays.

7.3.3   Royal Institute of British Architects (RIBA) work stages

The basic, very well-established, well-understood (and therefore well-used) RIBA work stages are interspaced with incidences of client approval, required before proceeding to the next stage. They can provide valuable insight into general international best practice, and provide invaluable client control and professional protection against scope 'creep' or changes in objective as the design and construction proceed.

Stage A   Inception site survey, Data collection, Feasibility
Stage B   Brief accommodation schedule, Design criteria
Stage C   Design concept, Design options, Cost plan
Stage D   Design development, Cost development
Stage E   Detail design documentation, Bills of quantity
Stage F   Submission documents and approvals
Stage G   Pre-qualification, Tender documentation, Tender adjudication
Stage H   Site handover and construction, Inspections, Queries and completion

The RIBA work stages described above have recently been amended to suit the purposes of BIM, but this list still demonstrates well the broad stages of the design process.

7.3.4   American Institute of Architects (AIA) work stages

The AIA has many chapters for individual states, each of which uses discrete data sheets, relevant to its own local working environments; however the main AIA website uses the following definition of **'how design works'**. The five basic stages are applicable to the design of individual homes but they uphold principles and relationships equally relevant to working on projects and programmes of all scales and complexity.

7.3.4.1 AIA five basic stages of design

- **Phase 1: Originate:** This first phase includes all of the discussions, thoughts and exploration that makes it apparent that something new must be created. The phase ends with your decision to move forward with a project.
- **Phase 2: Focus:** Here you define the project – its scope, features, purpose and functionality. This is the time to select an architect and architect agreement. With your architect, you develop and refine a vision for the project. Your architect leads you through a programming exercise to help you explore the needs of those who will live, work or play in the space you create. You identify the services you need from your architect, and the design team will begin to form a cohesive relationship and a shared concept for the final building.
- **Phase 3: Design:** Once the requirements of the project are determined, the design phase begins. Your architect gives shape to your vision through drawings and written specifications. Your input into this phase is vital, as you get the first glimpses, and then a more defined look, at how your building will take shape. It is important to establish a clear decision-making process with your architect during this phase. The design phase ends when you agree to the plans that will guide construction.
- **Phase 4: Build:** The contractor who will construct your building becomes the most active member of the team during this phase. Investments are made in materials, and timetables

are extremely important. Good communication within the project team is critical, as the need for changes often arises. This is typically the time of highest stress for the project owner. Your architect will discuss changes and options with you, and ensure that alterations are compatible with your vision for the project.

■ **Phase 5: Occupy:** This phase begins the day the project is up and running, and never really ends. During this phase your satisfaction with the project is determined. If you are turning over the project to others who will ultimately use it, good communication during that process is important. Your architect can help ensure that the terms of your building contract were met, and can use the experience of this project to inform future work, should you team together again. For these reasons, it is a good idea to maintain a relationship with your architect.

7.3.5 Simplified design work stages

However, in everyday practice, a more simplified five-stage design process can be used:

Stage 1   Accommodation schedule
Stage 2   Preliminary concept design
Stage 3   Schematic designs
Stage 4   Design approvals and documentation
Stage 5   Tenders

This format is often used internationally by less sophisticated client bodies in fast-track requests for proposal and design contracts; however it reduces the actual process to a minimum and should be seen as a symbolic representation of what is naturally a more complex undertaking in reality.

7.4. Design tools for programme management

As new technology emerges, designers and programme managers are learning to use the new tools creatively and practically. This often involves designers interacting with software designers directly during the software development phase and including programme management planning activities in the design concept stage.

In cases where the capability exists within design houses, designers routinely now modify and customise standard software programmes to suit specific project tasks or indeed to produce distinctive house styles, important for branding and destination definition in large programmes.

7.4.1 Parametric and algorithmic techniques for design in programme management

Contemporary digital design practice is in a state of rapid evolution. While architects have employed CAD systems for decades, only recently have two distinct and potent design sensibilities – parametric and algorithmic design – emerged. Nurtured by early architectural researchers and programmers operating in practice, these methodologies are now gaining widespread professional and academic acceptance.

7.4.2 Parametric design

'Parametric' is a term used in a variety of disciplines from mathematics through to design. Literally it means working within parameters of a defined range. Within the field of contemporary design, it refers broadly to the utilisation of parametric modelling software.

In contrast to standard software packages based on datum geometric objects, parametric software links dimensions and parameters to geometry, thereby allowing for the incremental adjustment of a part, thus affecting the whole assembly.

For example, as a point within a curve is repositioned the whole curve comes to realign itself. Parametric software therefore lends itself to curvilinear design.

However, it would be wrong to assume that parametric design is concerned primarily with form-making. On the contrary, parametric techniques afford the architect with new modes of efficiency compared to standard approaches, and new ways of coordinating the construction process (such as BIM or BIMM), as in the case of Digital Project, an architectural version of CATIA (computer-aided three dimensional interactive application) customised for the building industry by Gehry Technologies.

7.4.3   Algorithmic design

'Algorithmic' is a term that refers to the use of procedural techniques in solving design problems. Technically an algorithm is a simple instruction. It therefore relates as much to standard analogue design processes, as it does to digital design processes.

Within the field of digital design, however, it (algorithmic design) refers specifically to the use of scripting languages that allow the designer to step beyond the limitations of the user interface, and to design through the direct manipulation not of form but of code.

Typically algorithmic design would be performed through computer programming languages like Rhino Script, Maya Embedded Language (MEL), Visual Basic or 3DMaxScript. In contrast, due to the difficulty of programming, the applications Generative Components and Grasshopper bypass code with pictographic forms of automation.

Algorithmic design exploits the capacity of the computer to operate as a search engine, and perform tasks that would otherwise consume inordinate amounts of time. It therefore lends itself to optimisation and other tasks beyond the limitations of standard design constraints.

Together these two techniques are opening up a new field of possibilities for architectural practice. Most significantly, they have been developed and refined primarily in commercial practice and not in academia.

Since the late 1990s, these advances have coincided with the emergence of a number of digital research units within commercial practices. These in-house digital research units have been developed as a means of ensuring that the complex buildings of today are designed and constructed efficiently, on time and within budget (vitally important in programme management).

7.4.4   Building information modelling management (BIMM)

New computerised design technology to improve BIMM which emerged in 2000–2010 is rapidly bringing the traditional architect's role, including the newer urban designer role, which emerged in the 1980s, back into focus for larger-scale projects.

In this sense the engineering approach of past decades is rapidly being transformed by intuitive and highly responsive emotive and creative responses to solving new problems.

Designers are thinking and working in new ways with fully integrated use of new technologies and social media. Good, fresh public opinions and statistics make more relevant and resonant designs. Of increasing importance for large-scale urban programmes where public input is essential, crowdsourcing is used to supplement the traditional democratic decision-making process.

Besides the obvious instantaneous communication advantages of general social media such as Twitter and Facebook for programme management, crowdsourcing has emerged as a useful tool in targeting and testing public opinion, while eliciting a more active participation in large urban and infrastructure projects from residents and interested and affected parties.

This valuable and targeted information exchange in real-time can be invaluable to urban designers and lead design engineers as they develop new solutions for input into the overall programme. Ever wider uses of crowdsourcing are being discovered as the media, and the practice of this new form of technology, evolve and become more defined.

Crowdsourcing is the practice of obtaining needed services, ideas or content by soliciting contributions from a large group of people, and especially from an online community, rather than from traditional employees or suppliers. This process is often used to subdivide tedious work or to fund-raise start-up companies and charities, and can also occur offline. It combines the efforts of numerous self-identified volunteers or part-time workers, where each contributor on their own initiative adds a small portion to the greater result. The term 'crowdsourcing' is a portmanteau of 'crowd' and 'outsourcing'; it is distinguished from outsourcing in that the work comes from an undefined public rather than being commissioned from a specific, named group.

Crowdsourcing can involve division of labour for tedious tasks split to use crowd-based outsourcing, but it can also apply to specific requests, such as crowd funding, a broad-based competition, and a general search for answers, solutions, or even a missing person.

Even in developing countries where smart mobile phone usage is the main source of online communication, crowdsourcing can be invaluable.

Crowdsourcing tools and other new information transfer systems can increasingly realistically mimic and extend human ability, when combined creatively in multi-disciplinary teams with a knowledgeable contractor and client in a contractually aligned programme.

Parametric and algorithmic design with good BIM and crowdsourcing with a sound programme charter can frequently yield astonishingly good results in record-breaking time.

This new facility, together with paradoxical elements of detailed project design, can lead to general public incredulity and scepticism before the new solutions are properly understood and assimilated into community consciousness. New tools are only as good as the professionals wielding them.

However the benefits of good programme managers and designers using new tools will have very long-lasting benefits and synergies for vast new future populations, cities and the built and natural environments.

7.5. Urban programme management
7.5.1 Charter cities
A charter city is a city in which the governing system is defined by the city's own charter document rather than by state, provincial, regional or national laws. It is believed that by 2050, almost 70% of the world's estimated 10 billion inhabitants – more than the number of people living today – will be part of massive urban networks.

The current New York University (NYU) research and implementation programme on charter cities taking place currently extends partnering into exciting new territory for innovative funding of key areas of the world's megacities, now defined as having populations of at least 10–15 million.

So, with similar new design and economic tools we can begin to respond to the massive growth caused by influx of mobile populations approaching 20–25 million people per megacity in the very near future.

Programme management will take an increasingly important role in urbanisation, as multiple, interlinked urban projects require coordination, integration and delivery of good design to avoid unsustainable, soulless and antisocial urban environments.

Place making is naturally extended to the larger city and regional scale by new disciplines of place branding and destination marketing, which are increasingly part of the design process for major urban programme management work.

Cities and their mayors worldwide have realised the power of good design and sound programme management when applied to setting up an attractive destination for increasing visitors, revenue and the quality of urban life.

The city of Bilbao in the northern Basque region of Spain presents a well-known but nevertheless remarkable story of putting a relatively unknown provincial place squarely on the map for increased economic success by using good design on an urban scale.

7.5.2 Destination and place branding
The sculptural, titanium-clad Guggenheim museum in Bilbao by Frank Gehry has rapidly become a symbol of hope and regeneration for the city and for social and sustainable change in the Basque region as a whole. This is an excellent example of truly iconic architecture and its power to transform people's lives directly, but not in an isolated manner. The Guggenheim is the titanium crown of a well-dressed city with good new stations, transportation and interlinked urban spaces with fine food and distinct place making – nothing less than good 'genius loci' in the classical or Renaissance terms familiar to Alberti.

Behind the shimmering facades and soaring gallery spaces of the new Guggenheim lies an example of an early programme management success, which has cleverly integrated and interconnected good architecture with individual urban place making and landscape design to create new and desirable destination sites in Bilbao.

This was achieved by the programme managers introducing new sustainable urban infrastructure, including a revamped metro and tram transport system, modal interchange and new commuter stations using stringent programme economic and change controls.

Bilbao now welcomes and successfully hosts the new crowds of local, regional and international visitors together with the growing numbers of residential central city citizens, in memorable yet sustainable style.

How a region markets itself must be believable and true so that the actual experience matches the reality. As part of this marketing process you must first understand what is being offered, then decide what parts of this offering are attractive to relevant target audiences and then package this offering in a clear brand description for the region. All the subsequent marketing of the region should be consistent by all stakeholders so that maximum return on investment is achieved and that target audiences develop a clear understanding of the unique offer from that destination.

7.6. Case studies in programme management

7.6.1 Case study 1: BBC programming for the 2012 London Olympics

Mark Smith, programme manager, described thus the challenges faced by his team in helping to deliver the excellent BBC digital coverage of the games using new software:

> The complexity of delivering 24-hour live coverage of every event, with up to 24 streams of video, web pages for each of the 30 venues, 204 teams, 10,490 athletes and the results of every event, alongside coverage of the Torch Relay and 2012 cultural events, required a large team of product development staff working alongside key suppliers, over an 18 month period up to the Games.
> The team had significant experience in 'agile' website development using scrum and Kanban techniques, but had to define how this would work alongside the more traditional 'waterfall' development with an immovable deadline. The team were also working with partners to deliver the new technologies needed to deliver the 2,500 hours of video and data driving the dynamic creation of thousands of web pages, which added to the complexity of coordinating the programme of work.
> The programme organised the teams into two enabling work streams for the delivery of the dynamic streaming video and data infrastructure and the remaining six for audience-facing propositions, including the main Olympic sports pages, a 2012 portal and torch relay, connected TV apps, mobile browser and apps and the red button service. This enabled each team with their own editorial and technical leadership, business analysts, UX designers, developers and testers to build their own products, within a common architecture and API's to the shared data services.
> As well as dividing the work by team, the functionality to be delivered was prioritised by 'core, target and additional' scope, which helped to group features or 'stories' into releases, from which was scheduled the delivery of the highest impact areas first. Working with the editorial and business stakeholders, the team established the audience value of features in terms of their expected reach, impact and value for money, against the technical challenges of various implementation options.
> Detailed estimates of the development and testing of each story comprising a feature or task, was undertaken, as part of the sprint planning process. Once approved through the governance process, the team were able to schedule the releases and assign the development and testing resources accordingly.
> To ensure the team kept to schedule with its commitment to deliver all the core and target functionality, the prioritised backlog of stories was tracked using 'burn-down' charts, with the actual achieved velocity monitored at the end of each fortnightly sprint and remedial actions taken to re-establish track as required.

7.6.2 Case study 2: Olympic design, venue and infrastructure delivery

The 2012 London Olympics, with their successes which have been widely broadcast as detailed in the previous section, have proved to be a remarkable actual delivery of the required elements for the Games: the latter were delivered on time, at an appropriate quality, and have had an important legacy role with social and environmental impact in parts of East London.

This largely positive outcome has mainly been achieved simply by increasing urban connectivity by strategic transport, cycle and pedestrian links to the Thames Gateway area from well-established areas of the city within an overall Games and Legacy programme management exercise on a previously unprecedented level of complexity and national effort, perhaps exceeded only in theoretical wartime scenarios.

New destinations for community sport, residential and commercial opportunities have been realised as a result of new connections and successful urban place making. Arguably the long-term successes of such sustainable legacy projects go a long way to justifying some of the inevitable economic overruns and security risks encountered by the 2012 Olympic delivery team. This could be described as a balanced and active approach to programme management.

Principles of science, art, philosophy, economic and environmental sustainability and sheer collective team spirit and engineering are currently combining with IT and new technologies through inspired design leadership, to produce genuinely fresh and new solutions for modern communities.

Good design and, by extension, good programme management, when applied to large urban-scale programmes such as the London 2012 Olympics, can therefore sometimes be characterised as 'mysterious and un-scientific', particularly if their evolution is not clearly communicated widely and effectively by the lead designers to the public, the client, the programme manager, and the engineering and construction teams. Good design nevertheless remains absolutely fundamental to true originality and innovation, in successfully solving the complex, interrelated and inter-dependent problems encountered in major international programmes today.

Good design inevitably brings with it the initial 'shock of the new'. Good design introduces an element of the unknown and therefore a specific risk which must be managed steadily within the overall programme objectives in order for the many and exciting economic and sustainable benefits and synergies of the integrated design process to be fully realised, thereby creating a good design and programme management solution.

REFERENCES

American Institute of Architects (AIA) (2014) http://www.aia.org/contractdocs/aias077630 (accessed 10/10/2014).

BBC (2014) http://www.bbc.co.uk/blogs/bbcinternet/2012/06/interactive_video_player_launc.html (accessed 10/10/2014).

Programme Management in Construction
ISBN 978-0-7277-6014-2

Chapter 8
Programme management – the future

8.1. Programme management trends
8.1.1 International trends

Programme management will be the essential tool for the developed countries, for countries currently undergoing large developments such as China, India and other (mainly Far Eastern) countries, as well countries affected by catastrophic events such as earthquakes, tsunamis, unprecedented storms or wars.

Developed countries, where most of the infrastructure is already established, need sustained programmes to keep their facilities operational. These facilities will include the maintenance, facility management, upgrading and expansion of their structures and infrastructure in all areas, such as:

- transport including roads, bridges, tunnels and airports
- health facilities
- public buildings
- educational facilities such as schools, universities, training centres
- defence-related projects
- electricity and water supply establishments.

It is imperative that countries affected adversely by war embark on very large programmes to rebuild them in a relatively short time.

The world will see also large programmes in renewable energy and some large nuclear energy programmes to satisfy the population's thirst for energy and to compensate for diminishing oil supplies.

In this regard, due to the scarcity of resources and due to the ever-increasing world population, programme managers will have to look at very efficient, modern, complex and state-of-the-art techniques to build these programmes by a combination of modelling, highly charged optimisation models using new intelligent mathematical models, and a continuing reliance on geographic information systems (GIS) and BIM.

Also, clients including governments and in some instances local authorities and private companies will have to look for highly trained and knowledgeable managers conversant in non-traditional methods and familiar IT, mathematical programming, methodical decision making, conformity and communication skills.

Beginning with Jan Smut's post-Second World War vision of a 'League of Nations', the United Nations (UN) has proved to have a chequered history of success and failure particularly in international détente and conflict resolution. Today the UN is providing social improvement on a grand scale through its multiple programmes. Since 1966 and the inception of the United Nations Development Program (UNDP) the organisation has been grappling with serious global issues but from a proactive construction and developmental basis.

The UN's Millennium Development Goals (MDG) set ambitious targets distilled into eight goals for attainment by 2015, namely:

(a) Eradicate extreme poverty and hunger.
(b) Achieve universal primary education.
(c) Promote gender equality and empower women.
(d) Reduce child mortality.
(e) Improve maternal health.
(f) Combat HIV/AIDS, malaria and other diseases.
(g) Ensure environmental sustainability.
(h) Global partnership for development.

The eighth goal represents a vast and exciting challenge for programme managers and their organisations worldwide.

As a visionary global organisation the UN has gone even further. The World We Want (WWW) campaign is already launched for what happens after 2015:

> The new agenda has to be universal, reflecting the nature of both the post-2015 development agenda and the Rio +20 final document – a document stating that 'the Sustainable Development Goals (SDGs) should be global and universal in nature and applicable to all countries, whilst taking into account the different realities, skills and levels of development of each country and safeguarding national policies and priorities'.
>
> Pajin, 2014

8.1.2 National trends

At a national level, strategic infrastructure projects and forward planning involving multiple project streams, on a longer-term basis the issues become even more sharply into focus as programme risk, time constraints, responsibility and budgets all increase.

In democratic countries, being subject to the will of the people and subject to powerful national and global trends implies that the period of this 'planning window' will inevitably remain open for even shorter times. The time-sensitive project and programme management team will increasingly be in demand.

Programme management will play an increasingly vital role in urban programmes, especially in national event management which involves capital city urban regeneration projects such as the Olympics.

As rapid urbanisation is likely to continue unabated at national level in most countries in future, the pressure on programme management teams to keep pace will remain a strong motivator.

8.1.3 Regional trends

At regional level, investing in and maintaining modern systems, for GIS, BIM and other programme management tools such as decision-making procedures including conformity, optimisation modelling and visual instant communication tools, becomes more affordable, but consensus is more difficult to achieve regarding strategic planning over a wide area of multiple authorities.

Stakeholder matrix management will therefore become an increasingly important task of the programme management team as they receive, analyse and re-distribute communications from multiple sources.

Regional destination planning and branding programmes are also likely to be more economical when shared and they will therefore increase the complexity of the overall regional programme management strategy as visitors and resident communities cluster closer around regional air, coach and rail terminals.

8.1.4   Local trends

Rural and urban local authorities around the world are increasingly under pressure to deliver more sustainable resources and more effective solutions for modern communities' need for programmes. These programmes can be smaller in size but more complex and more numerous than large programmes. A local authority's need to refurbish a number of training grounds, parks, substations and roads can be a very complex matter especially if these are to be built on a stringent budget with limited staff, and contractors not conversant in programme management.

As the authorities also grapple with climate change and rapid urbanisation, good local programme management combined with GIS as a more powerful tool will be increasingly needed to monitor, record and regulate the multiple, inter-linked projects required to deal with complex problems and to bring costs and risks down to acceptable levels.

Wider action areas such as transport, flooding catchments, water supply, waste disposal together with regular social services, emergency medical and food relief, vaccination and epidemic control, can all be captured effectively by GIS, be broken down into work elements, and thence be captured by the local authority in their strategic planning and programme management processes.

BIM, as described in earlier chapters, can be used to simplify regulatory and approvals procedures for authorities. New systems require substantial investment in new IT systems and training of key personnel as well as ongoing upgrading and maintenance costs. This capital cost will rapidly be justified by improved economic and strategic performance once new systems are established and being used well.

Depending on the geographic, budgetary and population factors of individual local authorities, sharing of combined GIS and programme management resources at a regional level will, in the view of the authors, rapidly become a trend.

8.2.   Future challenges, tools, and new techniques for design in programme management

With the challenges that programme management faces, the development of new tools and techniques in respect of the programme management design process is essential. Some of these techniques and possible future trends are considered below.

8.2.1   Parametric design trends

Parametric design is a recent trend in computer-aided architectural design. An effective parametric building model manages object data at the component level, but more importantly allows information about relationships between all of the components, views and annotations in the model to be shown. This is the strength of a parametric building model. It records, presents and manages relationships between the parts of the building no matter where they occur in the building.

Around the world, more and more amazing architecture is being designed and contracted successfully through parametric design methods. However, discussions about the creative process in parametric design are limited. Using parametric design methods, architects can rapidly generate design alternatives, which in turn may promote reflections and re-examinations of design problems. This process may help designers to broaden their understanding of design problems and foster their creativity.

8.2.2   Algorithmic design trends

In terms of computers, algorithms have been commonly defined as 'instructions for completing a task'. Perhaps a more accurate description would be that algorithms are patterns for completing a task in an efficient way.

Algorithms are good for creating very complex geometries with small amounts of data. They work well with the way nature constructs using cellular components. The fractal (or self-similar) nature that you see in trees and leaf veins and arteries is due to this.

But the code here is embedded in the cell itself and cells organise themselves to create complex forms based on relatively simple code. The important aspect of this is that this repetitiveness is part of the build process operating at a cellular level and not at a blueprint level.

The bulk of architectural mathematical theory was formed in pre-modern days, well before the geometric basis of nature was understood. However, algorithms can be used particularly for creating interesting surface patterns.

With the present developments under way, it appears clear that we are about to witness a new explosion of algorithmic form. This, initially, will not necessarily be in architecture but in 3D printed products. 3D printing is ideally suited for algorithmic form because in 3D printing complexity comes for free – whereas in architecture complexity comes with a cost.

However, programme managers, architects and engineers are yet to figure out how the next layer of complexity can give birth to building rooms, ventilation, views and visual delight. Until that time, we should stop making glorious connections to old-world mathematics. A new world of form is now being born out of a new world of mathematical and biological understanding, and we are all yet to come to terms with it.

8.2.3   Building information modelling management (BIMM) trends

BIM stands for 'building information modelling' or 'building information management'; and because of this, sometimes the acronym BIMM is used, standing for 'building information modelling and management'.

At its most basic level, BIM is a computerised process which is used to design, understand and demonstrate the key physical and functional characteristics of a building on a 'virtual' computerised model basis. BIM therefore provides the opportunity to concurrently design and visualise the building in 3D.

At its more advanced levels, BIM is the use of a computer software model to simulate the construction and operation of a building. The resulting model, a building information model, is a digital representation of the building, from which views and data appropriate to various users' needs can be extracted and analysed to generate information that can be used to make decisions and improve the process both for the building's delivery and for the entire lifecycle of the building.

8.3.   Future trends in programme management
8.3.1   Future city programmes
Engineering and construction professionals' understanding of the urban environment, whether in theory or practice, stands at a turning point. Cities all over the world face complex and rapidly evolving challenges, such as climate change, global migration flows, transnational governance demands, financial volatility and expanding social inequalities.

Existing political, socio-economic and technical lock-in and path dependencies involve high switching costs that reduce innovation in decision making. Addressing these challenges requires ingenuity and versatility, whether in policy making, investment decisions or everyday livelihoods. Yet mainstream understandings of cities and how to transform them often derive from rigid concepts, models and practices about the urban environment.

The present aims are towards rethinking the city as a flexible and dynamic space that responds better to evolving circumstances. The focus on the 'flexible city' presents scholars, policymakers, investors and the public at large with an inherently interdisciplinary perspective and multi-faceted approach to question contemporary concepts, methodologies and policies towards urban change.

8.3.2   Resource programmes
Resource management is the efficient and effective deployment and allocation of an organisation's resources when and where they are needed. Such resources may include financial resources, inventory, human skills, production resources or information technology. Resource management includes planning, allocating and scheduling resources to tasks, which typically include manpower, machines, money and materials. Resource management has an impact on schedules and budgets as well as on resource levelling and smoothing.

In order to effectively manage resources, organisations must have data on resource demands forecasted by time period into the future, on the resource configurations that will be required to meet those demands, and on the supply of resources, again forecasted into the future. Forecasts should be as far into the future as is reasonable.

8.3.3   The changing programme paradigm
As the price of resources increases, organisations, including clients, consultants and contractors, will need to find new and better methods of building programmes to offset costs while remaining competitive in the marketplace or to satisfy the stakeholders and the public. The recent rise in the prices for fuel, with accompanying rise in cost of other resources, and the scarcity of some resources, such as cement, steel, wood, aluminium and basic elements in construction such as gravel, copper and iron, will become inevitable to programme leaders in the construction.

The increased cost of programme resources must be factored into the programme plan and resource allocation for passing along to the client. The senior programme management team must find ways to absorb the extra cost of resources, or raise the price for the projects resulting from programmes.

Current programme management procedures and practices continue to find improvements that can save on the consumption of resources, but there is still room for improvement. Programme managers are frequently assigned to programmes for the purpose of delivering buildings and structures that work with the emphasis on schedule and technical performance. Programme management literature focuses on programme managers' effectiveness and on functional managers' efficiency, with delivery of projects implied as constituting such effectiveness.

Under current conditions, programme managers cannot always be both efficient and effective.

8.4. Trends that dictate the need for change in programme management procedures

Programme management effectiveness and efficiency do not occur by accident or through ad hoc, generic and simple procedures. Continuous change in a random manner cannot bring about the best results.

There is always an unexpected wastage and planning slippage in building programmes. These areas of inefficiency include rework of parts of the programme because the initial job was done incorrectly, installation methods were not followed correctly, and materials were initially misused. Programme delivery may have resulted in lost time, more cost, and not the quality expected.

More staff may have been required than were planned, and more materials consumed than were necessary, which was driving up cost of constructing the programmes – though it is also recognised that programme management has matured significantly over the past 25 years.

Programme managers will be expected to use fewer resources to reduce the programme cost – and perhaps the only way to do this is to eliminate waste from the programme.

Programme inefficiency can be compensated for in the following ways:

- minimising staff time used for the programme
- better allocation of resources such as materials and equipment
- construction methods being optimised
- less idle time while decisions are pending
- utilising the right tools for activities
- optimal plan for implementation.

Engineering must coordinate all programme external interfaces with the stakeholders and report this to the programme manager, but should not make changes to the interfaces without specific approval from the programme manager. The planning process can identify both weaknesses in the conceptual approach and paths that are not feasible options.

A technically competent individual who does not have the leadership, negotiation and other requisite behavioural skills has a decreased chance of successfully managing a programme.

Understanding the contextual aspects of a programme provides, for example, the information to support decision making. Programme managers are better performers when they are competent in the technical, contextual and behavioural categories of programme management.

8.5. Possible improvements in programme management in the future

The roles of senior managers and programme managers will be defined to clearly show each team member's responsibility in planning. Programme managers will have improved planning skills that allow guiding a team to develop a plan that includes:

- a clear mission or purpose statement that describes what the programme is to accomplish
- a comprehensive list of facts and assumptions that are related to accomplishing the mission
- clear objectives that support the mission or purpose statement
- a well-defined schedule that clearly describes the activities, their durations, and allocated resources
- the requisite information collection, formatting and dissemination procedures
- a description of the periodic review practices
- a programme close-out procedure that includes delivery of each project.

Other areas of programmes such as risk and procurement will also be required.

Programme leaders must review and approve the programme plan against objective criteria. This procedure may entail a review and comment on the programme plan by peers or functional managers with a stake in the programme. Significant is how closely the programme is to be monitored and what the urgency for the product is, which relates to the priority assigned to the programme.

A list of potential responsibilities includes the following:

- Select only those programmes that contribute to the organisation's mission.
- Select programmes that fit into the strategic portfolio of programme work.
- Select and assign only competent programme managers to lead programmes.
- Approve or reject programme plans that either meet or fail to meet the organisation's policy.
- Review and comment upon programme progress – redirect programme direction, when necessary.
- Terminate unproductive programmes at the earliest possible time.

Organisations will conduct rehearsals of programmes based on approved plans. These rehearsals will comprise two or three days of 'war gaming' the programme to identify strengths and weaknesses while exercising the core staff. Benefits of the rehearsal will include the following:

- Permit the core team to work together to build relationships and enhance communication among the core team.
- Stress the implementation and close-out phases of the programme to determine whether the plan has the right level of detail – not too much and not too little.
- Highlight weaknesses in the plan for fixing, prior to actual programme implementation.
- Smooth out control procedures and information collection points to ensure adequate control over the programme.

- Compress the programme duration by having a core team that is fully prepared to conduct programme work.
- Reduce cost for lost time in rework and decision making by the programme manager and senior managers.

Defining roles and responsibilities for programmes is key to each person meeting his or her responsibilities. When these responsibilities are undefined, stakeholders develop their respective tasks that may or may not align with the organisation's goals. Organisational competence is not optimised unless there is good alignment with the organisation's business purpose.

8.6. Conclusion

In conclusion, there is danger in the comfort zone for businesses when programme leaders become satisfied with the status quo. Change is coming and clients, contractors and consultants need to be ready to meet these challenges.

Often it takes a crisis to motivate leaders to make change – and that is only because they must adapt to survive. Anticipating and planning for change is difficult because many people feel comfortable with the existing environment. Future-looking leaders plan for the future and implement effective changes that meet the new situations that emerge.

More improvements will be necessary through research on current practices and assessment of the worth of new practices. Improvements may be made internally to programme management practices, but the greatest improvement may be changes to organisational management of programmes. Organisational competence in programme management can perhaps have more dramatic results on programme success than can improvements in programme management practices.

Organisational effectiveness in employing programmes to support programme and portfolio goals makes a major contribution to the success of programmes. Assessing the organisational structure and business design can identify areas that need improvement to optimise programme selection, support, and management in the organisation. This would result in better chances for the programme manager to deliver projects that meet the clients' needs.

Organisational practices can either select the right programme to meet strategic goals or select one that is in *contretemps* with the programme's purpose. Programme selection is perhaps the most important aspect of senior management in an organisation. Also, how the programme is supported in actual practice makes a valuable contribution to programme success. The winners will be those anticipating change in time to adjust to the new environment through new and better practices. Will programme managers struggle to meet greater productivity goals or will they manage in a professional, measured way? While the time for designing change for the future is now, may the best programme leaders decide what is needed to meet these challenges.

REFERENCE

Pajin L (2014) Millennium Development Goals to Sustainable Development Goals. http://www.undp.org/content/undp/en/home/ourperspective/ourperspectivearticles/2014/04/28/de-los-objetivos-del-milenio-a-los-de-desarrollo-sostenible-leire-paj-n/ (accessed 08/10/2014).

Programme Management in Construction
ISBN 978-0-7277-6014-2

ICE Publishing: All rights reserved
http://dx.doi.org/10.1680/pmic.60142.139

Index